Sue & Mary
Energy Efficient Building
Oxford Brookes University

10/07/01

Energy-efficient buildings in India

— Suchita Rajat

Energy-efficient buildings in India

Foreword
R K Pachauri, Director, TERI

Editor
Mili Majumdar

Tata Energy Research Institute

Ministry of Non-conventional Energy Sources

© Ministry of Non-conventional Energy Sources and Tata Energy Research Institute 2001

This book has been prepared under MNES project no.15/11/97-SEG(ST). No part of this publication may be reproduced or transmitted in any form, without written permission of MNES and TERI.

ISBN 81-85419-82-5

Edited by
Mili Majumdar

Technical contribution by
Anil Misra, TERI
Pradeep Kumar, TERI
Bibek Bandyopadhyay, MNES

Cover design by
Rasik Varsani

Published by
Tata Energy Research Institute
Darbari Seth Block, Habitat Place
Lodhi Road, New Delhi – 110 003, India
Telephone +91 11 468 2100, 468 2111
Fax +91 11 468 2144, 468 2145
E-mail mailbox@teri.res.in
Web www.teriin.org

and

Ministry of Non-conventional Energy Sources
Government of India
Block No.14, CGO Complex
Lodhi Road, New Delhi – 110 003

Disclaimer
The contents of the book reflect the technical and other features of the projects as provided by the respective project architects. The MNES (Ministry of Non-conventional Energy Sources) and TERI (Tata Energy Research Institute) do not assume any responsibility for the authenticity of the design, costs, performance data, and any other information contained in the book. The MNES and TERI will also not be liable for any consequences arising out of use of any information or data contained in the book.

Printed by
Thomson Press (India) Limited
B-315, Okhla Phase - I
New Delhi – 110 020

Contents

vii   Foreword

ix   Preface

xi   Acknowledgements

1   Energy efficiency in architecture: an overview of design concepts and architectural interventions

Climatic zone: cold and cloudy

21   Himurja office building, Shimla
25   Himachal Pradesh State Co-operative Bank, Shimla
30   Residence for Mohini Mullick, Bhowali, Nainital
33   MLA Hostel, Shimla

Climatic zone: cold and sunny

41   Degree College and Hill Council Complex, Leh
45   Airport and staff housing colony, Kargil
49   LEDeG Trainees' Hostel, Leh
55   Sarai for Tabo Gompa, Spiti

Climatic zone: composite

59   Residence for Madhu and Anirudh, Panchkula
62   PEDA office complex, Chandigarh
66   Bidani House, Faridabad
69   Transport Corporation of India Ltd, Gurgaon
76   SOS Tibetan Children's Village, Rajpur, Dehradun
80   Redevelopment of property at Civil Lines, Delhi
86   Integrated Rural Energy Programme Training Centre, Delhi
91   Tapasya Block (Phase 1), Sri Aurobindo Ashram, New Delhi
97   Residence for Sudha and Atam Kumar, Delhi
102   Residence for Neelam and Ashok Saigal, Gurgaon
106   Dilwara Bagh, Country House for Reena and Ravi Nath, Gurgaon
111   RETREAT: Resource Efficient TERI Retreat for Environmental Awareness and Training, Gurgaon
119   Water and Land Management Institute, Bhopal
124   Baptist Church, Chandigarh
128   Solar Energy Centre, Gual Pahari, Gurgaon
134   National Media Centre Co-operative Housing Scheme, Gurgaon
138   American Institute of Indian Studies, Gurgaon

Climatic zone: hot and dry

- 145 Indian Institute of Health Management Research, Jaipur
- 151 Sangath – an architect's studio, Ahmedabad
- 155 Torrent Research Centre, Ahmedabad
- 161 Residence for Mahendra Patel, Ahmedabad
- 166 Solar passive hostel, Jodhpur

Climatic zone: moderate

- 173 Residence for Mary Mathew, Bangalore
- 177 TERI office building-cum-guest house, Bangalore

Climatic zone: warm and humid

- 185 Nisha's Play School, Goa
- 189 Office building of the West Bengal Renewable Energy Development Agency, Kolkata
- 195 Office-cum-laboratory for the West Bengal Pollution Control Board, Kolkata
- 201 Silent Valley, Kalasa
- 206 Vikas Apartments, Auroville
- 210 La Cuisine Solaire, Auroville
- 214 Kindergarten School, Auroville
- 218 Visitors' Centre, Auroville
- 223 Computer Maintenance Corporation House, Mumbai

Appendices

- 231 Commonly used software packages in energy-efficient building analysis and design
- 233 List of institutions / organizations, architects, and scientists working on energy-efficient buildings
- 234 Financial incentives by the Ministry of Non-conventional Energy Sources for the promotion of solar passive architecture
- 236 Explanation of the temperature check charts provided in the book
- 237 Energy-efficient glazings
- 240 Products and services offered by various companies

- 249 Index

Foreword

What we once called a 'house for all seasons' has now been christened 'climate-conscious' or 'bioclimatic' or 'energy-efficient' or 'sustainable' architecture. Of late, impressive terms like these have found recognition in the vocabulary of contemporary architects and urban designers. With enough reason, of course, as sustainability in urban spaces increasingly becomes an imperative, particularly as urban habitats keep extending their footprints on the earth.

Sustainable architecture aims to create environment-friendly and energy-efficient buildings. This entails actively harnessing renewable natural resources like solar energy and utilizing materials that cause the least possible damage to the global commons—water, soil, forests, and air.

Increased development of housing and commercial buildings has imposed immense pressure on our dwindling energy sources and other vital resources like water, thus aggravating the already rampant process of environmental degradation. The Ninth Plan of the Government of India has taken a lead by stressing the development and improvement of urban areas as economically efficient, socially equitable, and environmentally sustainable entities.

With the emphasis now on sustainable habitats as a key solution to growing urban concerns, which assume a global dimension, this book provides a very timely insight into the context, techniques, and benefits of energy-efficient building. The examples provided here encapsulate various combinations of response to climatic conditions; improvized blends of traditional and innovative building techniques; material selection; implementation of energy-efficient building systems; and use of renewable energy systems as viable alternatives to powering buildings through conventional energy sources.

This well-designed and informative book will serve the dual purpose of (1) educating laypersons about the multiple benefits of building in tune with nature and (2) reinvigorating the zeal of professional designers, builders, and planners to impart to their projects a bond with the earth far greater than any other aesthetic value.

(R K Pachauri)
Director, TERI

Preface

A penny saved is a penny earned, they said. So with joules of energy! With recent exponential increases in energy pricing, the formerly neglected or underestimated concept of energy conservation has swiftly assumed great significance and potential in cutting costs and promoting economic development, especially in a developing-country scenario.

Reckless and unrestrained urbanization, with its haphazard buildings, has bulldozed over the valuable natural resources of energy, water, and ground cover, thereby greatly hampering the critical process of eco-friendly habitat development.

However, it is not too late to retrace the steps. The resource crunch confronting the energy supply sector can still be alleviated by designing and developing future buildings on the sound concepts of energy efficiency and sustainability.

Energy efficiency in buildings can be achieved through a multi-pronged approach involving adoption of bioclimatic architectural principles responsive to the climate of the particular location; use of materials with low embodied energy; reduction of transportation energy; incorporation of efficient structural design; implementation of energy-efficient building systems; and effective utilization of renewable energy sources to power the building.

India is quite a challenge in this sense. N K Bansal and Gernot Minke (1988), in their book entitled *Climatic Zones and Rural Housing in India*, have classified Indian climate into six major zones: cold and sunny, cold and cloudy, warm and humid, hot and dry, composite, and moderate. Translation of bioclimatic architectural design in the Indian context, therefore, provides a plethora of experiences and success stories to learn from. Several buildings have come up, fully or partially adopting the above approach to design.

The book you hold is the result of a comprehensive survey of a few such buildings. Our objectives include:

- creating awareness of designing energy-efficient building envelopes by taking advantage of the climatic conditions of a particular region
- highlight resource-efficient building practices in India
- promote the application of efficient lighting and HVAC (heating, ventilation, and air-conditioning) systems to reduce energy demand
- advocate the application of renewable energy systems
- provide performance studies of the buildings in the form of user feedback or monitoring results (wherever available)
- quantify both the savings achieved by energy-efficient buildings as against conventional ones and the incremental costs (wherever available)
- compiling a database of energy-efficient building products and services, available software, and reference books

A survey questionnaire, prepared in consultation with subject experts, was circulated to several architects. Advertisements were posted in leading architectural magazines requesting details of energy-efficient building projects. Several submissions were received in response. Review visits to numerous sites and dialogues eliciting user feedback on performances of the building yielded the 41 projects that have been covered in this book.

Spanning all climatic zones of India, these projects are selected on the basis of their energy-efficient approach to design—be it adoption of low-energy construction material and techniques, innovative use of passive solar architectural principles, or use of renewable energy systems. Projects not included here are the ones that used some stand-alone renewable energy gadgets/devices without having an integrated approach to the whole design.

This book will prove to be of interest and of benefit to practising architects, building designers and scientists, engineers, urban planners, architecture students, municipal authorities, policy makers, and concerned citizens. We expect this book to serve not only as a handy reference document but also as a source of inspiration to correct our building concepts and practices.

Editor

Acknowledgements

The MNES (Ministry of Non-conventional Energy Sources), Government of India, has taken important initiatives to promote solar passive architecture in India. TERI is grateful to the MNES (Ministry of Non-conventional Energy Sources) for providing financial and technical assistance for the production of this book. TERI is especially grateful to Dr E V R Sastry, Adviser, MNES, and Dr Bibek Bandyopadhyay, Director, MNES, for their valuable inputs and suggestions.

There have been a number of people who have contributed to this project and TERI would like to thank everyone for their time, effort, and help in making this book a success.

- Dr R K Pachauri, Director, TERI, for his constant support, encouragement, and for providing all the infrastructural support
- Dr Ajay Mathur, former Dean of the Energy–Environment Technology Division, TERI, and Dr V V N Kishore, Senior Fellow, TERI, for their guidance in structuring the contents
- Mr Somnath Bhattacharjee, Dean of the Energy–Environment Technology Division, TERI, for his guidance
- Air Cmde M M Joshi, Chief, Administrative Services, TERI, for providing administrative support
- Ms Rajeshwari Prakash Menon, architect and journalist, for editorial assistance
- Dr N K Bansal, Professor, Centre of Energy Studies, Indian Institute of Technology, New Delhi for his valuable suggestions
- Mr Anil Misra, Consultant, TERI, and Mr Pradeep Kumar, Fellow, TERI, for providing technical inputs and assistance in the project work
- Mr K P Eashwar for editing and managing the production of the book
- Mr Saif Hyder Hasan, Dr Vani J Shankar, Mr Kaushik Das Gupta, and Ms Mudita Chauhan for editorial assistance
- Mr R Ajith Kumar for graphics, design, and typesetting
- Mr Rasik Varsani, University of Reading, UK, for conceptualizing and designing the cover page
- Mr R K Joshi for making the cover page
- Mr Girish Kumar for graphic assistance
- Mr P K Jayanthan for indexing the book
- Mr Jitendra Bakshi for sourcing advertisements
- Mr I I Jose, Ms Tina Alawadi, and Mr A Joseph for secretarial assistance
- Mr T Radhakrishnan for supervising the production of this book
- All the advertisers for supporting the book

Energy efficiency in architecture: an overview of design concepts and architectural interventions

Buildings, as they are designed and used today, contribute to serious environmental problems because of excessive consumption of energy and other natural resources. The close connection between energy use in buildings and environmental damage arises because energy-intensive solutions sought to construct a building and meet its demands for heating, cooling, ventilation, and lighting cause severe depletion of invaluable environmental resources.

However, buildings can be designed to meet the occupant's need for thermal and visual comfort at reduced levels of energy and resources consumption. Energy resource efficiency in new constructions can be effected by adopting an integrated approach to building design. The primary steps in this approach are listed below.

- *Incorporate solar passive techniques in a building design to minimize load on conventional systems (heating, cooling, ventilation, and lighting)* Passive systems provide thermal and visual comfort by using natural energy sources and sinks, e.g. solar radiation, outside air, sky, wet surfaces, vegetation, and internal gains. Energy flows in these systems are by natural means such as radiation, conduction, and convection with minimal or no use of mechanical means. The solar passive systems vary from one climate to the other. For example, in a cold climate, an architect's aim would be to design a building in such a way that solar gains are maximized, but in a hot climate, the architect's primary aim would be to reduce solar gains, and maximize natural ventilation.

- *Design energy-efficient lighting and HVAC (heating, ventilation, and air-conditioning) systems* Once the passive solar architectural concepts are applied to a design, the load on conventional systems (HVAC and lighting) is reduced. Further, energy conservation is possible by judicious design of the artificial lighting and HVAC system using energy-efficient equipment, controls, and operation strategies.

- *Use renewable energy systems (solar photovoltaic systems / solar water heating systems) to meet a part of building load* The pressure on the earth's non-renewable resources can be alleviated by judicious use of earth's renewable resources, i.e. solar energy. Use of solar energy for meeting electrical needs of a building can further reduce consumption of conventional forms of energy.

- *Use low energy materials and methods of construction and reduce transportation energy* An architect should also aim at efficient structural design, reduced use of transportation energy and high energy building material (glass, steel, etc.) and use of low energy building materials.

Thus, in brief, an energy-efficient building balances all aspects of energy use in a building – lighting, space-conditioning, and ventilation – by providing an optimized mix of passive solar design strategies, energy-efficient equipment, and renewable sources of energy. Use of materials with low embodied energy also forms a major component in energy-efficient building designs.

The book carries 41 case studies on energy- and resource-efficient architectural projects in India. Each project highlights the energy-efficiency measures adopted by the architects. The projects have been classified climate-wise. The thermal performances of a selected number of buildings have also been presented. The incremental costs for incorporation of energy-efficiency measures to buildings have been included wherever such data were available.

The passive architectural techniques that have been adopted by the architects have been discussed in this paper.

Architects can achieve energy efficiency in the buildings they design by studying the macro and microclimate of the site, applying bioclimatic architectural principles to combat the adverse conditions, and taking advantage of the desirable conditions. A few common design elements that directly or indirectly affect thermal comfort conditions and thereby the energy consumption in a building are listed below.

- Landscaping
- Ratio of built form to open spaces
- Location of water bodies
- Orientation
- Planform
- Building envelope and fenestration.

However, in extreme climatic conditions, one cannot achieve comfortable indoor conditions by these design considerations alone. There are certain tested and established concepts, which, if applied to a design in such climatic conditions, are able to largely satisfy the thermal comfort criterion. These are classified as advanced passive solar techniques. The two broad categories of advanced concepts are (1) passive heating concepts (direct gain system, indirect gain system, sunspaces, etc.) and (2) passive cooling concepts (evaporative cooling, ventilation, wind tower, earth-air tunnel, etc.).

The commonly considered design elements for achieving lower energy consumption in a building are discussed below.

Common design elements

Landscaping

Landscaping is an important element in altering the microclimate of a place. Proper landscaping reduces direct sun from striking and heating up building surfaces. It prevents reflected light carrying heat into a building from the ground or other surfaces. Landscaping creates different airflow patterns and can be used to direct or divert the wind advantageously by causing a pressure difference. Additionally, the shade created by trees and the effect of grass and shrubs reduce air temperatures adjoining the building and provide evaporative cooling. Properly designed roof gardens help to reduce heat loads in a building. A study shows that the ambient air under a tree adjacent to the wall is about 2 °C to 2.5 °C lower than that for unshaded areas (Bansal, Hauser, and Minke 1994).

Trees are the primary elements of an energy-conserving landscape. Climatic requirements govern the type of trees to be planted. Planting deciduous trees on the southern side of a building is beneficial in a composite climate. Deciduous plants such as mulberry or Champa cut off direct sun during summer, and as these trees shed leaves in winter, they allow the sun to heat the buildings in winter. This landscaping strategy has been adopted to shade the southern side of the RETREAT (Resource Efficient TERI Retreat for Environmental Awareness and Training) building of TERI (see page no. 111)

▲ The RETREAT building has deciduous trees on the south side to cut off summer gains. These trees shed leaves during winter so that winter solar gains are not cut off. Wind breaks are provided in the north and north-east to protect from the winter winds

Building form/surface-to-volume ratio

The volume of space inside a building that needs to be heated or cooled and its relationship with the area of the envelope enclosing the volume affect the thermal performance of the building. This parameter, known as the S/V (surface-to-volume) ratio, is determined by the building form. For any given building volume, the more compact the shape, the less wasteful it is in gaining/losing heat. Hence, in hot and dry regions and cold climates, buildings are compact in form with a low S/V ratio to reduce heat gain and losses, respectively. Also, the building form determines the airflow pattern around the building, directly affecting its ventilation. The depth of a building also determines the requirements for artificial lighting—greater the depth, higher the need for artificial lighting.

Location of water bodies

Water is a good modifier of microclimate. It takes up a large amount of heat in evaporation and causes significant cooling especially in a hot and dry climate. In humid climates, water should be avoided as it adds to humidity. Water has been used effectively as a modifier of microclimate in the WALMI (Water and Land Management Institute) building complex in Bhopal (see page no. 119).

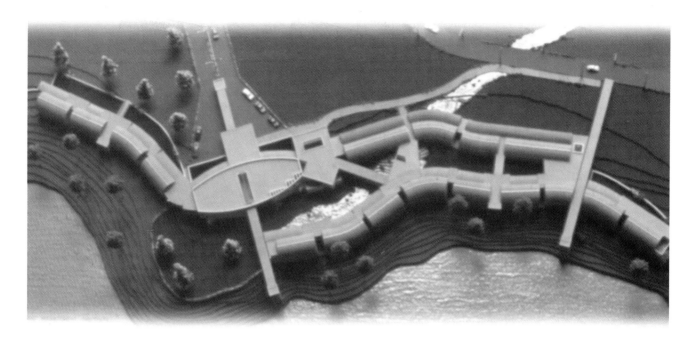

A view of the WALMI building. The flowing form overlooks a water body which has been used to advantage for modification of the microclimate

Orientation

Building orientation is a significant design consideration, mainly with regard to solar radiation and wind. In predominantly cold regions, buildings should be oriented to maximize solar gain; the reverse is advisable for hot regions. In regions where seasonal changes are very pronounced, both the situations may arise periodically. For a cold climate, an orientation slightly east of south is favoured (especially 15 degrees east of south), as this exposes the unit to more morning than afternoon sun and enables the house to begin to heat during the day.

This has been amply demonstrated in the MLA Hostel building at Shimla (see page no. 33). Similarly, wind can be desirable or undesirable. Quite often, a compromise is required between sun and wind orientations. With careful design, shading and deflecting devices can be incorporated to exclude the sun or redirect it into the building, just as wind can be diverted or directed to the extent desired.

Building envelope and fenestration

The building envelope and its components are key determinants of the amount of heat gain and loss and wind that enters inside. The primary elements affecting the performance of a building envelope are

- materials and construction techniques
- roof
- walls
- fenestration and shading
- finishes.

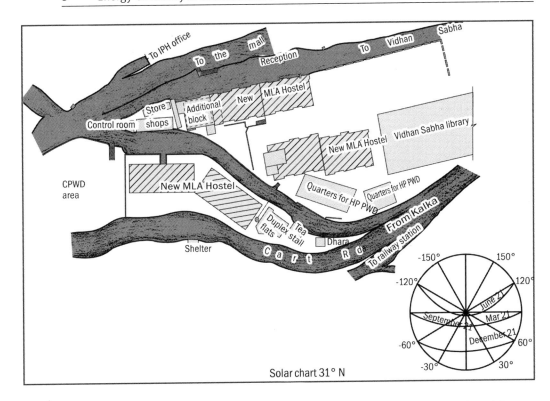

▲ The building blocks in the MLA hostel, Shimla, located in the cold and cloudy zone, are oriented due south (±15 degrees) for direct solar gain. They are spaced apart so as to eliminate shadows of one building over the other, even for the longer winter shadows. It was proposed that all bedrooms be south-facing to avail of the benefit of south exposure

Materials and construction techniques

Material with low embodied energy

Choice of building materials is important in reducing the energy content of buildings. Strain on conventional energy can be reduced by use of low-energy materials, efficient structural design, and reduction in transportation energy.

The choice of materials also helps to maximize indoor comfort. Use of materials and components with low embodied energy has been demonstrated in various buildings in Auroville. The Visitors' Centre at Auroville uses innovative materials and construction techniques to reduce embodied energy of the building and attain the desired comfort conditions conducive to the warm and humid climate.

Thermal insulation

Insulation is of great value when a building requires mechanical heating or cooling insulation helps reduce the space-conditioning loads. Location of insulation and its optimum thickness are important. In hot climates, insulation is placed on the outer face (facing exterior) of the wall so that thermal mass of the wall is weakly coupled with the external source and strongly coupled with the interior (Bansal, Hauser, and Minke 1994).

Use of 40-mm thick expanded polystyrene insulation on walls and vermiculite concrete insulation on the roof has brought down space-conditioning loads of the RETREAT building by about 15%.

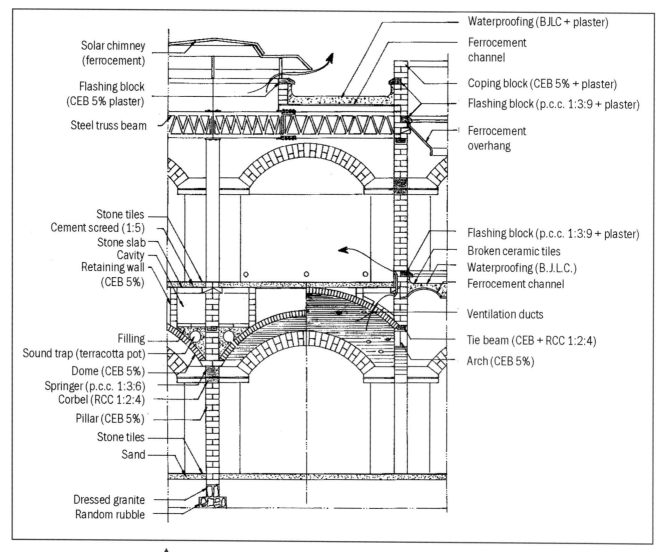

▲ Sectional details showing use of alternative construction techniques. Ferrocement solar chimney and ventilation duct in corbelled arch demonstrates use of passive solar ventilation techniques

Table 1 Energy contents of commonly used building materials

Building Elements/materials		KWh/m^3 (in situ)
Cement concrete	1:5:10	402
Lime concrete with brick ballast	1:4:8	1522 (80% in brick)
Brick masonry	1:5	676
Brick masonry	1:4	709
Random rubble masonry	1:4	267
Stabilized mud with 6% lime	–	197
Stabilized mud with 10% lime	–	320
RCC roof (10 cm)	–	$174/m^2$
Stone slabs in RCC joists	–	$132/m^2$
Cement plaster	1:4	$20.65/m^2$
Cement plaster	1:6	$15.09/m^2$
Lime surkhi	1:4	$11.05/m^2$

Source Gupta (1994)

Roof

The roof receives significant solar radiation and plays an important role in heat gain/losses, daylighting, and ventilation. Depending on the climatic needs, proper roof treatment is essential. In a hot region, the roof should have enough insulating properties to minimize heat gains. A few roof protection methods are as follows.

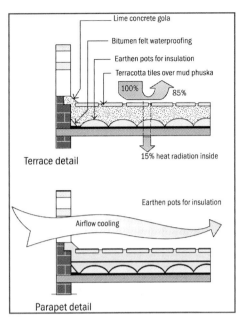

▲ Roof details showing use of earthern pots for roof insulation

◀ Broken China mosaic can be used as topmost layer in roof for reflection of incident radiation

- A cover of deciduous plants or creepers can be provided. Evaporation from leaf surfaces will keep the rooms cool.
- The entire roof surface can be covered with inverted earthen pots. It is also an insulating cover of still air over the roof.
- A removable cover is an effective roof-shading device. This can be mounted close to the roof in the day and can be rolled up to permit radiative cooling at night. The upper surface of the canvas should be painted white to minimize the radiation absorbed by the canvas and consequent conductive heat gain through it.
- Effective roof insulation can be provided by using vermiculite concrete. This has been used in the RETREAT building at Gual Pahari (near New Delhi) and has reduced roof conduction by 60%.

▲

The skylight on the roof of the West Bengal Renewable Energy Development Agency office building provides natural light for inner office spaces and is connected to the office spaces for inducing ventilation by stack effect

The roof can also be used advantageously for effective ventilation and daylighting by incorporating vents and skylights. This has been demonstrated effectively in the recently constructed office building of the WBREDA (West Bengal Renewable Energy Development Agency) in Kolkata.

Walls

Walls are a major part of the building envelope and receive large amounts of solar radiation. The heat storage capacity and heat conduction property of walls are key to meeting desired thermal comfort conditions. The wall thickness, material, and finishes can be chosen based on the heating and cooling needs of the building. Appropriate thermal insulation and air cavities in walls reduce heat transmission into the building, which is the primary aim in a hot region.

Air cavities

Air cavities within walls or an attic space in the roof ceiling combination reduce the solar heat gain factor, thereby reducing space-conditioning loads. The performance improves if the void is ventilated. Heat is transmitted through the air cavity by convection and radiation. A cavity represents a resistance, which is not proportional to its thickness. For a thickness >20 mm, the resistance to heat flow remains nearly constant. Ventilated air does not reduce radiative heat transfer from roof to ceiling. The radiative component of heat transfer may be reduced by using low emissivity or high reflective coating (e.g. aluminium foil) on either surface facing the cavity. With aluminium foil attached to the top of ceiling, the resistance for downward heat flow increases to about 0.7 m^2k/W, compared to 0.21 m^2k/W in the absence of the foil (Bansal, Hauser, and Minke 1994).

Fenestration and shading

Of all the elements in the building envelope, windows and other glazed areas are most vulnerable to heat gain or losses. Proper location, sizing, and detailing of windows and shading form an important part of bioclimatic design as they help to keep the sun and wind out of a building or allow them when needed.

The location of openings for ventilation is determined by prevalent wind direction. Openings at higher levels naturally aid in venting out hot air. Size, shape, and orientation of openings moderate air velocity and flow in the room; a small inlet and a large outlet increase the velocity and distribution of airflow through the room. When possible, the house should be so positioned on the site that it takes advantage of prevailing winds. The prevailing wind direction is from the south/south-east during summer. The recommendations in IS:3362-1977 code of practices for natural ventilation of residential buildings (first revision) should be satisfied in the design of windows for lighting and ventilation. There should be sufficient air motion in hot-humid and warm-humid climates. In such areas, fans are essential to provide comfortable air motion indoors. Fenestrations having 15%–20% of floor area are found adequate for both ventilation and daylighting in hot and dry, and hot and humid regions.

Natural light is also admitted into a building through glazed openings. Thus, fenestration design is primarily governed by requirements of heat gain and loss, ventilation, and daylighting. The important components of a window that govern these are the glazing systems and shading devices.

Glazing systems

Before recent innovations in glass, films, and coatings, a typical residential window with one or two layers of glazing allowed roughly 75%–85% of the solar energy to enter a building. Internal shading devices such as curtains or blinds could reflect back some of that energy outside the building. The weak thermal characteristics of windows became a prime target for research and development in an attempt to control the indoor temperature of buildings. A detailed write-up on energy-efficient glazing system is provided in Appendix V.

Windows admit direct solar radiation and hence promote heat gain. This is desirable in cold climates, but is critical in hot climates. The window size should be kept minimum in hot and dry regions. For example, in Ahmedabad, if glazing is taken as 10% instead of 20% of the floor area, then the number of uncomfortable hours in a year can be reduced by as much as 35% (Nayak, Hazra, and Prajapati 1999).

Shading devices

Heat gain through windows is determined by the overall heat loss coefficient U-value (W/m^{2}K) and the solar energy gain factor, and is much higher as compared to that through solid wall. Shading devices for windows and walls thus moderate heat gains into the building. In a low-rise residential building in Ahmedabad (hot and dry climate), shading a window by a horizontal 0.76-m deep *chhajja* can reduce the maximum room temperature by 4.6 °C (from 47.7 to 43.1 °C). Moreover, the number of uncomfortable hours in a year with temperatures exceeding 30 °C can be reduced by 14% ((Nayak, Hazra, and Prajapati 1999).

Shading devices are of various types (Bansal, Hauser, and Minke 1994).
1. Moveable opaque (roller blind, curtains, etc.) can be highly effective in reducing solar gains but eliminate view and impede air movement.
2. Louvres (adjustable or fixed) affect the view and air movement to some degree.
3. Fixed overhangs.

Relative advantages and disadvantages of these shading devices are given below.

Moveable blinds or curtains
- Block the transmission of solar radiation through glazed windows, especially on the east and west walls
- In hot and dry climates, when ambient air is hotter than room air, they help to reduce convective heat gain
- In warm, humid climates, where the airflow is desirable, they impede ventilation
- For air-conditioned buildings, where the flow of outside air is to be blocked, they can reduce cooling load.

Overhangs and louvres
- Block that part of the sky through which sunlight passes
- Overhangs on south-oriented windows provide effective shading from the high-altitude sun
- An extended roof shades the entire north or south wall from the noon sun

The office building for the West Bengal Pollution Control Board is a landmark of energy and resource conscious architecture in this region. Efficient planning and carefully designed shading devices, fenestration design and efficient lighting design has brought about 40% energy savings over a conventional building of similar size and function. This picture shows the east facade with inclined louvres to cut off solar gains

- East and west openings need much bigger overhangs, which may not be possible and can be achieved by porticos or verandahs on these sides or by specially designed louvres to suit the building requirements.

The scientific design of fenestration and shading devices in the West Bengal Pollution Control Board building has brought down the projected energy consumption substantially (TERI 1996).

Finishes

The external finish of a surface determines the amount of heat absorbed or rejected by it. For example, a smooth and light colour surface reflects more light and heat in comparison to a dark colour surface. Lighter colour surfaces have higher emissivity and should be ideally used for warm climate.

Advanced passive heating techniques

Advanced passive heating techniques are used by architects in building design to achieve thermal comfort conditions in cold climate. Passive solar heating systems can be broadly classified into direct gain systems and indirect gain systems

Direct gain

Direct gain is the most common passive solar system. In this system, sunlight enters rooms through windows, warming the interior space. The glazing system is generally located on the southern side to receive maximum sunlight during winter (in the northern hemisphere). The glazing system is usually double-glazed, with insulating curtains to reduce heat loss during night. South-facing glass admits solar energy into the building, where it strikes thermal storage materials such as floors or walls made of adobe, brick, concrete, stone, or water. The direct gain system uses 60%–75% of solar energy striking the windows. The interior thermal mass tempers the intensity of heat during the day by absorbing heat. At night, the thermal mass radiates heat into the living space, thus warming the spaces.

Direct gain can be achieved by various forms of openings such as clerestories and skylight windows designed for the required heating. Direct gain systems have been used for day-use rooms by architect Sanjay Prakash in the residence for Mohini Mullick at Bhowali. The user is extremely satisfied with the thermal performance of the direct gain system in this residence.

The direct gain system of the Bhowali house. The picture highlights the fully glazed walls for the day-use rooms from inside

Direct gain systems have some limitations. They cause large temperature savings (typically 10 °C) because of large variations in input of solar energy. In direct gain systems, the interior of the house receives direct sunlight, which results in the degradation of the interior furnishings, etc. However, being relatively simple to construct and inexpensive, they are by far the most common systems used worldwide.

Indirect gain system

In an indirect gain system, thermal mass is located between the sun and the living space. The thermal mass absorbs the sunlight that strikes it and transfers it to the living space. The indirect gain system uses 30%–45% of the sun's energy striking the glass adjoining the thermal mass. A few commonly used indirect gain systems are discussed below.

Trombe wall

A Trombe wall is a thermally massive wall with vents provided at the top and bottom. It may be made of concrete, masonry, adobe, and is usually located on the southern side (in the northern hemisphere) of a building in order to maximize solar gains. The outer surface of the wall is usually painted black for maximizing absorption and the wall is directly placed behind glazing with an air gap in between.

Solar radiation is absorbed by the wall during the day and stored as sensible heat. The air in the space between the glazing and the wall gets heated up and enters the living spaces by convection through the vents. Cool air from the rooms replaces this air, thus setting up a convection current. The vents are closed during night, and heat stored in the wall during the day heats up the living space by conduction and radiation.

Trombe walls have been extensively used in the cold regions of Leh. Various forms of Trombe walls have been tried and tested in the LEDeG Hostel at Leh (read *LEDeG Trainees' Hostel* (pp. 49–54) for their advantages).

It is worth noting that in buildings with thermal storage walls, indoor temperature can be maintained at about 15 °C when the outside temperature is as low as –11 °C (Mazria 1979).

Generally, thickness of the storage wall is between 200 mm and 450 mm, the air gap between the wall and glazing is 50–150 mm, and the total area of each row of vent is about one per cent of the storage wall area (Levy, Evans, and Gardstein 1983). The Trombe wall should be adequately shaded for reducing summer gains.

Water wall

Water walls are based on the same principle as that for Trombe walls, except that they employ water as the thermal storage material. A water wall is a thermal storage wall made up of drums of water stacked up behind glazing. It is usually painted black to increase heat absorption. It is more effective in reducing temperature swings, but the time lag is less.

Heat transfer through water walls is much faster than that for Trombe walls. Therefore, the distribution of heat needs to be controlled if it is not immediately required for heating the building. Buildings that work during daytime, such as schools and offices, benefit from the rapid heat transfer in the water wall. Overheating during summer may be prevented by using suitable shading devices.

Roof-based air heating system

In this technique, incident solar radiation is trapped by the roof and is used for heating interior spaces. In the northern hemisphere, the system usually consists of an inclined south-facing glazing and a north-sloping insulated surface on the roof. Between the roof and the insulation, an air pocket is formed, which is heated by solar radiation. A moveable insulation can be used to reduce heat loss through glazed panes during nights. There could be variations in detailing of roof air heating systems. In the Himachal Pradesh State Cooperative Bank building, south-glazing is in the form of solar collectors warming the air and a blower fan circulating the air to the interior spaces.

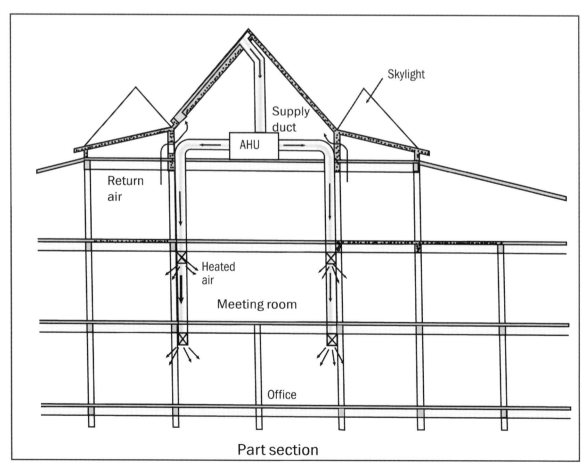

▲ Roof-based air heating system for winter heating in the Himachal Pradesh State Cooperative Bank building

Sunspaces

A sunspace or solarium is the combination of direct and indirect gain systems. Solar radiation heats up the sunspace directly, which, in turn, heats up the living space (separated from the sunspace by a mass wall) by convection and conduction through the mass wall. In the northern hemisphere, the basic requirements of buildings heated by sunspace are (1) a glazed south-facing collector space attached yet separated from the building and (2) living space separated from the sunspace by a thermal storage wall. Sunspaces may be used as winter gardens adjacent to the living space. The Himurja building in Shimla has a well-designed solarium on the south wall to maximize solar gain.

Advanced passive cooling techniques

Before the turn of the century, buildings were designed to take advantage of daily temperature variations, convective breeze, shading, evaporative cooling, and radiation cooling. However, with a thoughtless imitation of the west, these concepts took a back seat and buildings became energy guzzlers. Today, with high energy costs and growing environmental concerns, many of these simpler techniques are once again becoming attractive. Passive cooling systems rely on natural heat-sinks to remove heat from the building. They derive cooling directly from evaporation, convection, and radiation without using any intermediate electrical devices. All passive cooling strategies rely on daily changes in temperature and relative humidity. The applicability of each system depends on the climatic conditions.

The relatively simple techniques that can be adopted to provide natural cooling in the building have been elaborated earlier.

These design strategies reduce heat gains to internal spaces. This section briefly elaborates the passive techniques that aid heat loss from the building by convection, radiation, and evaporation, or by using storage capacity of surrounding spaces, e.g. earth berming.

Ventilation

Outdoor breezes create air movement through the house interior by the 'push-pull' effect of positive air pressure on the windward side and negative pressure (suction) on the leeward side. Good natural ventilation requires

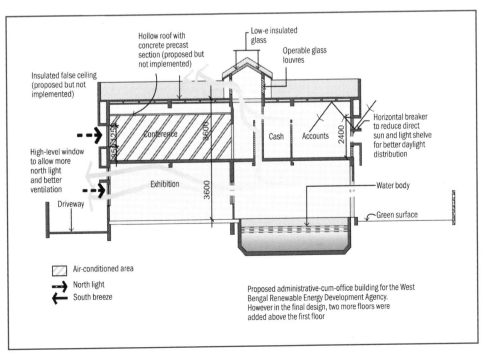

Section of the WBREDA office building showing ventilation strategies

locating openings in opposite pressure zones. Also, designers often choose to enhance natural ventilation using tall spaces called stacks in buildings. With openings near the top of stacks, warm air can escape whereas cooler air enters the building from openings near the ground. Ventilation by creating stacks has been effectively used in the office building of the West Bengal Renewable Energy Development Agency in Kolkata (see page no. 189). Located in a warm humid climate, induced ventilation was a primary design strategy for this building.

Innovative ventilation strategies by use of building-integrated solar chimneys have been used in Sudha and Atam Kumar's residence in the composite climate of New Delhi (see page no. 97).

The windows, as discussed earlier, play a dominant role in inducing indoor ventilation due to wind forces. Other passive cooling techniques that induce indoor natural ventilation and are used by architects to achieve passive cooling are discussed below.

Building-integrated solar chimney in Sudha and Atam Kumar's residence in New Delhi for effective ventilation especially during humid season. The inset shows a closer view of the chimney top

Wind tower
In a wind tower, the hot air enters the tower through the openings in the tower, gets cooled, and thus becomes heavier and sinks down. The inlet and outlet of rooms induce cool air movement. In the presence of wind, air is cooled more effectively and flows faster down the tower and into the living area. After a whole day of air exchanges, the tower becomes warm in the evenings. During the night, cooler ambient air comes in contact with the bottom of the tower through the rooms. The tower walls absorb heat during daytime and release it at night, warming the cool night air in the tower. Warm air moves up, creating an upward draft, and draws cool night air through the doors and windows into the building. The system works effectively in hot and dry climates where diurnal variations are high. The Jodhpur Hostel, designed by Dr Vinod Gupta, uses wind tower for summer cooling.

A wind tower works well for individual units not for multi-storeyed apartments. In dense urban areas, the wind tower has to be long enough to be able to catch enough air. Also protection from driving rain is difficult.

Wind tower in Jodhpur Hostel to catch favourable cool wind from south-west for passive cooling

Courtyard effects

Due to incident solar radiation in a courtyard, the air gets warmer and rises. Cool air from the ground level flows through the louvred openings of rooms surrounding a courtyard, thus producing air flow.

At night, the warm roof surfaces get cooled by convection and radiation. If this heat exchange reduces roof surface temperature to wet bulb temperature of air, condensation of atmospheric moisture occurs on the roof and the gain due to condensation limits further cooling.

If the roof surfaces are sloped towards the internal courtyard, the cooled air sinks into the court and enters the living space through low-level openings, gets warmed up, and leaves the room through higher-level openings. However, care should be taken that the courtyard does not receive intense solar radiation, which would lead to conduction and radiation heat gains into the building. Intensive solar radiation in the courtyard also produces immense glare.

Earth air tunnels

Daily and annual temperature fluctuations decrease with the increase in depth below the ground surface. At a depth of about 4 m below ground, the temperature inside the earth remains nearly constant round the year and is nearly equal to the annual average temperature of the place. A tunnel in the form of a pipe or otherwise embedded at a depth of about 4 m below the ground will acquire the same temperature as the surrounding earth at its surface and, therefore, the ambient air ventilated though this tunnel will get cooled in summer and warmed in winter and this air can be used for cooling in summer and heating in winter.

Earth air tunnel has been used in the composite climate of Gurgaon in the RETREAT building. The living quarters (the south block of RETREAT) are maintained at comfortable temperatures (approximately between 20 °C and 30 °C) round the year by the earth air tunnel system, supplemented, whenever required, with a system of absorption chillers powered by liquefied natural gas during monsoons and with an air washer during dry summer. However, the cooler air underground needs to be circulated in the living space. Each room in the south block has a 'solar chimney'; warm air rises and escapes through the chimney, which creates an air current for the cooler air from the underground tunnels to replace the warm air. Two blowers installed in the tunnels speed up the process. The same mechanism supplies warm air from the tunnel during winter (for details, see page nos. 111–118).

Evaporative cooling

Evaporative cooling lowers indoor air temperature by evaporating water. It is effective in hot and dry climate where the atmospheric humidity is low. In evaporative cooling, the sensible heat of air is used to evaporate water, thereby cooling the air, which, in turn, cools the living space of the building. Increase in contact between water and air increases the rate of evaporation.

The presence of a water body such as a pond, lake, and sea near the building or a fountain in a courtyard can provide a cooling effect. The most commonly used system is a desert cooler, which comprises water, evaporative pads, a fan, and pump. Evaporative cooling has been tried as a roof-top installation at the Solar Energy Centre, Gurgaon. However, the system has now become defunct due to poor water supply in the area.

Passive downdraught cooling

Evaporative cooling has been used for many centuries in parts of the Middle East, notably Iran and Turkey. In this system, wind catchers guide outside air over water-filled pots, inducing evaporation and causing a significant drop in temperature before the air enters the interior. Such wind catchers become primary elements of the architectural form also. Passive downdraught evaporative cooling is particularly effective in hot and dry climates. It has been used to effectively cool the Torrent Research Centre in Ahmedabad.

▲ Passive downdraught cooling has been successfully used at the Torrent Research Centre, Ahmedabad. The wind catchers for the system are the predominant architectural elements in this building

Daylighting

Daylighting has a major effect on the appearance of space and can have considerable energy-efficiency implications, if used properly. Its variability and subtlety is pleasing to the occupants in contrast to the relatively monotonous environment produced by artificial light. It helps to create optimum working conditions by bringing out the natural contrast and colour of objects. The presence of natural light can bring a sense of well being and awareness of the wider environment. Daylighting is important particularly in commercial and other non-domestic buildings that function during the day. Integration of daylighting with artificial lighting brings about considerable savings in energy consumption.

A good daylighting system has a number of elements, most of which must be incorporated into the building design at an early stage. This can be achieved by considering the following in relation to the incidence of daylight on the building.

- Orientation, space organization, and geometry of the space to be lit
- Location, form, and dimensions of the fenestrations through which daylight will enter
- Location and surface properties of internal partitions that affect daylight distribution by reflection
- Location, form, and dimensions of shading devices that provide protection from excessive light and glare
- Light and thermal characteristics of the glazing materials.

Daylight integration is an important aspect of energy-efficient building design, and most of the case studies covered in this book have innovative daylighting strategies.

Conclusion

This books contains 41 case studies of energy- and resource-efficient architecture, which have used one or a combination of the above concepts and techniques. In addition to the above, many of the projects have adopted innovative daylighting strategies. Use of energy-efficient lighting and space-conditioning strategies are the primary strengths of some buildings.

In the present era of growing environmental concerns, these case studies would inspire an architect to design and create a better tomorrow.

References

Bansal N K, Hauser G, and Minke G. 1994
Passive Building Design: A Handbook of Natural Climatic Control
Amsterdam: Elsevier Science B.V.

Gupta C L. 1994
Energy contents of building materials for India
Paper presented at the *Green Architecture Festival*
Organized by the Indian Institute of Architects, Nasik Centre, Nasik, India, February 1994
[data courtesy Sanjay P, Rakesh Ahuja, and Geeta V]

Levy M E, Evans D, and Gardstein C. 1983
The Passive Solar Construction Handbook
Pennsylvania: Rodale Press

Mazria E. 1979
The Passive Solar Energy Book
Pennsylvania: Rodale Press

Nayak J K, Hazra R, and Prajapati R. 1999
Manual on solar passive architecture
Mumbai: Energy Systems Engineering, Indian Institute of Technology, Mumbai;
New Delhi: Solar Energy Centre, Ministry of Non-conventional Energy Sources, Government of India. 105 pp.

SP:41 (S&T). 1987
Handbook on Functional Requirements of Buildings
New Delhi: Bureau of Indian Standards

TERI. 1996
Design review of West Bengal Pollution Control Board building at Salt Lake, Calcutta
Submitted to the West Bengal Pollution Control Board
New Delhi: Tata Energy Research Institute [TERI report 1995RT65]

Climatic zone: cold and cloudy

Himurja office building, Shimla

Himachal Pradesh State Co-operative Bank, Shimla

Residence for Mohini Mullick, Bhowali, Nainital

MLA Hostel, Shimla

Himurja office building, Shimla

Architects Arvind Krishan and Kunal Jain

Office building for Himachal Pradesh Energy Development Agency with active and passive solar retrofits

Editor's remarks

The Himurja office building is a unique example of energy-efficient architecture demonstrating application of several passive and active solar interventions in an existing building. It also highlights the higher cost of retrofitting an existing building to make it efficient as compared to design stage implementation of similar measures.

Located in Shimla at an altitude of about 2000 m above mean sea level in the 'middle' Himalayas, the Himurja building has been designed and built in a climate zone that is cold and cloudy albeit with a fair number of sunny days. The sharply sloping site provides a classical situation in a hilly urban context for a building within a large commercial complex that thus sits against the mountain for the lower three floors and inevitably has a 'deep plan'.

A south-west view of the office building showing specially designed sunspaces for maximizing solar gains in winter

The climate

Shimla, while lying in the 'cold and cloudy' climate zone, has a fairly long winter – October to February end – with a severe cold spell of about two months minimum DBT (dry bulb temperature) – 3 °C with short wet periods in winter. While summer (May and June) is pleasant with a maximum DBT of 28 °C, monsoon period (July and August) has a high level of precipitation with high humidity (maximum relative humidity 85%). Intervening periods have a milder climate.

Design response

The climate requires buildings to be heated almost throughout the year (a requirement that becomes quite necessary in winters). A detailed analysis of the site revealed good southern exposure and a good possibility of additional solar heat gain from the western exposure.

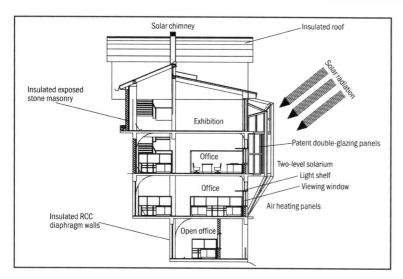

▲ Section through the building showing solar access through specially designed solarium and air heating panels

The demand on the building, therefore, is for heating for most part of the year, and for good ventilation during summer, and for good daylight distribution throughout the year.

Daylighting and heating

The plan of the building and its three-dimensional form allow maximum penetration of sun, maximizing both solar heat gain and daylight. While heat gain is maximized, its absorption in the judiciously designed thermal mass provides heat in the spaces throughout the diurnal cycle. Air heating panels designed as an integral part of the southern wall panels provide effective heat gain through a close connective loop. Distribution of heat gain in the entire building is achieved through a connective loop utilizing the stairwell as a means of distributing heated air through the principle of buoyancy. Since solar heat gain raises the internal ambient temperature above the comfort range in summers even though the outside conditions are quite comfortable, ventilation is an effective strategy for summers for

A closer view of the solarium ▶

Plan showing daylighting levels at regular intervals ▶

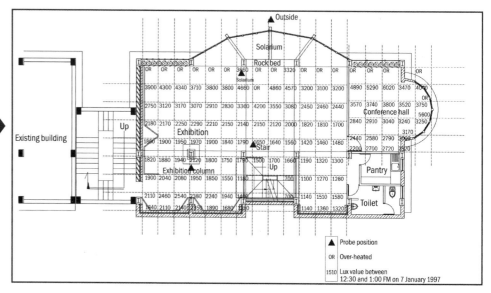

dissipating internal heat build-up. To optimize ventilation, the connective loop is coupled with solar chimneys designed as an integral part of the roof. Specially designed solarium (sunspace) is built as an integral part of southern wall to maximize heat gain.

Distribution of daylight in spaces is achieved through a careful integration of window and light shelves. Light reflected off the light shelves is distributed into the deep plan of the building by designing a ceiling profile that provides effective reflectivity.

A view of curved ceiling with glass blocks to distribute daylight and roof-mounted solar water heating system

Insulation and window design

Good insulation of 5-cm thick glass wool and minimum fenestration (only in toilets) on northern exposure prevents heat loss. Infiltration losses are minimized through weather-proofed (with no thermal bridges) hard plastic windows. Double-glazing helps control heat loss from glazing without creating any internal condensation.

Renewable energy systems

The photovoltaic system of 1.5 kWp meets the energy demand for lighting whenever required. Artificial lighting is seldom required (except during dark sky conditions sometimes in winters) in the south-oriented spaces, which are well daylit during working hours. Roof-mounted solar hot water system (1000 litre per day) has been used in the building. The water is circulated through radiators for space heating specially in the northern spaces.

Performance

It has been monitored both for thermal performance and daylight distribution during active use and found that it provides a comfortable working environment.

In the month of January, the coldest part of the year, the building does not require any auxiliary heaters. The building was monitored in the month of January 2001. The inside temperatures were recorded between 18 °C and 28 °C corresponding to ambient temperatures of 9–15 °C. Ventilation effectively generates a fresh ambient condition within the entire building. Daylight distribution allows good availability of daylight in the working zone. Even in the rear spaces of the deep plan building, the level of daylight is 150 Lux. Consequently, the office building requires no electrical energy for heating, lighting in daytime, and for hot water, which has been achieved through a solar water heating system. Ventilation is achieved through the connective loop activated by air buoyancy. This building can, therefore, be described as a 'zero energy consuming' building during the daytime—the normal hours of operation.

User feedback

- Excellent thermal conditions exist inside the building, specially it is comfortable in winters in almost all areas.
- The top-most floor has the problem of overheating in summer.

At a glance

Project details

Location Shimla, Himachal Pradesh
Building type Office building
Climate Cold and cloudy
Architects Arvind Krishan and Kunal Jain
Owner/client Himachal Pradesh Energy Development Agency (Himurja)
Year of completion 1997
Built-up area 635 m²
Cost The initial cost of the building was estimated at Rs 7 million (without incorporation of passive or active solar measures). Additional amount of Rs 1.3 million was incurred due to incorporation of passive and active solar measures. Thus there was an increase of 18.6% over initial cost by adoption of these measures. The high additional cost is attributed to the fact that solar systems were retrofitted onto an already constructed building.

Design features

- Air heating panels designed as an integral part of the south wall provide effective heat gain. Distribution of heat gain in the building through a connective loop that utilizes the stairwell as a means of distributing heated air
- Double-glazed windows with proper sealing to minimize infiltration
- Insulated RCC diaphragm walls on the north to prevent heat loss
- Solar chimney
- Specially designed solarium on south for heat gain
- Careful integration of windows and light shelves ensures effective daylight distribution
- Solar water heating system and solar photovoltaic system

Temperature check for Shimla

(see Appendix IV)

Cold	Cool	Comfortable	Warm	Hot	
	●				January
	●				February
	●				March
	●				April
		●			May
		●			June
		●			July
		●			August
	●				September
	●				October
	●				November
	●				December

Add the checks

	8	4		

Multiply by 8.33% for % of year

Heating	67
Comfortable	33

Himachal Pradesh State Co-operative Bank, Shimla

Architect Ashok B Lall

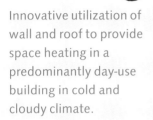

Editor's remarks

Innovative utilization of wall and roof to provide space heating in a predominantly day-use building in cold and cloudy climate.

Innovative combination of solar passive and active systems for a predominantly day-use building to cut down heating needs during winters

This multi-storeyed RCC (reinforced concrete cement) structure is located in Shimla. Oriented at 10 degree west of south, the building has a long narrow profile in plan. The narrow south face has access to light and air from the adjacent lane while the east face has no access to light and air as it abuts another building.

Since Shimla experiences good sunshine during winters and plenty of sun falls on façade of the building, a considerable amount of solar heat could be collected from this façade. The bulk of energy consumption in the building was on two fronts, one for lighting the deeper spaces in the centre of the building and the other for heating the building during winters.

South-facing heat collector wall with a dark wall behind. The heated air from the surface of this Trombe wall is drawn out at the top of the staircase tower and conducted through ducts to the central workspaces.

Energy-saving strategies

Design temperature

The design temperature was taken to be 18 °C for winter, considering that it is customary to be warmly dressed while working.

Limiting conditioning to work-spaces only

It was decided that passages, lobbies, stairs, toilets, and stores need not be conditioned. Only occupied office work spaces are to be conditioned.

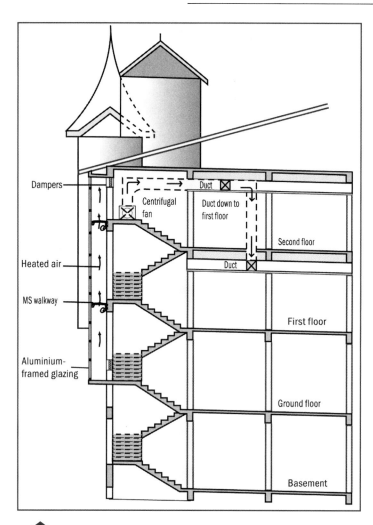

Section of the building with heat collector wall highlighting flow of heated air through the building

Sunspaces

The balconies adjacent to rooms on the southern face are converted into glazed sunspaces. These act as winter gardens and heat collectors for the adjacent room thereby reducing glare, enhancing views, increasing comfort, and also livening up the main façade.

Heat collector wall

The remaining south-facing surface was designed as a heat collecting wall by placing a continuous glass façade on the outer face of the building. The wall is to be clad with dark-coloured slate and ceramic mural, which would be visible. The mural would get lit at night as an advertisement of the bank. The heated air from the surface of this wall is drawn out at the top of the staircase tower. The cool air from the main banking hall gets heated through convection by the wall.

Roof collector

A roof top solar collector has been installed. This has been angled at 45 degrees to receive the winter sun. The collector warms up air, which is circulated into the spaces with a blower. An insulated air handling room is located below the attic space.

Details of air heating panels on roof for space heating
An electric back-up heating system is linked with the solar passive heating system. This electric bank is thermostatically controlled and switches on progressively on demand—during mornings and prolonged cloudy weather.

Roof-based air heating system

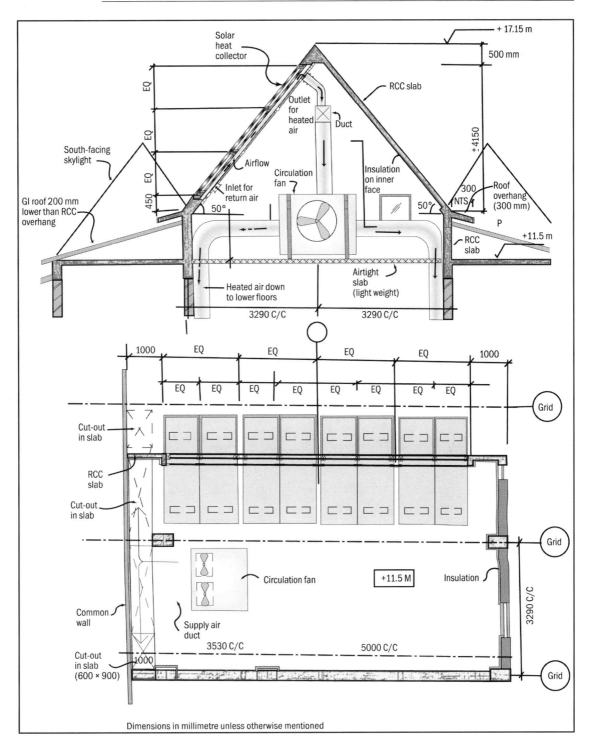

▲ Rooftop detail for solar collector (plan and section)

The system is switched on at 9.00 a.m., electrical back-up covers till about 11.00 a.m. when solar heating takes over. The system is designed for 10% fresh air supply, return air goes back to solar air panels, and supplies hot air to three halls.

Fenestration

Double-glazing and tight-fitting openable panels for windows have been installed for circulation and infiltration control. On the top floor, the north and west walls and the ceiling are insulated.

Daylighting and artificial illumination

Two light wells against east-side of the building were suggested at strategic locations, for the dual purpose of ensuring sufficient daylight into the banking hall and main office floor, and return air ducting. However, these wells could not be provided due to the high priority accorded to floor space utilization.

The artificial electric light system has been planned in a manner such that the lights could be switched on as a supplement to the available daylight. This arrangement would not incur any extra cost.

User feedback

The system provides thermal comfort during winters (saving significant amount of electricity and money). The interior of the building is quite warm in winters. Temperature rise of about 5–10 °C is achieved through incorporation of solar passive features.

Ample daylight is available due to sunspaces. The performance of the building has been monitored during winters. An annual savings of about 200 000 rupees in space heating has been achieved in one heating season of 6 months by adoption of energy-saving techniques (Energy cost of about 160 000 rupees was incurred in one heating season for running the back-up system as against an estimated energy cost of about 370 000 rupees if conventional heating system was used).

At a glance

Design features

- Sunspaces on the southern side
- Solar wall on the southern side
- Solar heat collector on roof-top with duct system for supply to various rooms
- Double-glazed windows
- Air-lock lobby at the main entrance

Project details

Location Mall Road, Shimla, Himachal Pradesh
Building type Office building
Client/Owner Himachal Pradesh Co-operative Bank
Architect Ashok B Lall
Climate Cold and cloudy
Local Architect C L Gupta
Energy consultant S S Chandel, Principal Scientific Officer and Coordinator, Solar House Action Plan Himachal Pradesh, State Council for Science, Technology, and Environment.
Year of start/completion 1995–1998
Built-up area 1650 m^2 (about 35% is heated by solar air heating system)
Total area of solar air heating panels 38 m^2
Cost of entire system Rs 1.1 million (includes AHU, electrical back-up, blower, ducting controls)
Electrical back-up 3 × 15 kW (in 3 stages)
Blower 4000 cfm (constant speed)
Brief specifications The external walls are 23-cm thick masonry construction with 5-cm thick glass wool insulation. The total window area is about 155 m^2 which are double-glazed and openable. The roofing is made of corrugated galvanized iron sheeting
Total building cost Rs 22 million (including solar passive and active features). The initial cost of the bank building without incorporation of passive solar measures was Rs 12 666/m^2, which was increased by Rs 680/m^2 to Rs 13 346/m^2 thus resulting in 5.6% increase in cost due to incorporation of passive solar measures

Temperature check for Shimla

(see Appendix IV)

Cold	Cool	Comfortable	Warm	Hot	
	•				January
	•				February
	•				March
	•				April
		•			May
		•			June
		•			July
		•			August
	•				September
	•				October
	•				November
	•				December

Add the checks

	8	4		

Multiply by 8.33% for % of year

Heating	67
Comfortable	33

Residence for Mohini Mullick, Bhowali, Nainital

Editor's remarks

This small family cottage is a unique example of traditional hill architecture that maximize the use of solar energy to meet needs of thermal comfort, water heating, and food warming. Use of local materials for construction reduces transportation energy.

Architect Sanjay Prakash

Traditional hill architecture maximizing the use of solar energy for meeting energy needs

Architect Sanjay Prakash believes that minimum damage to hill ecology is done when you build as little as possible. For the cottage, she planned to live in after retirement, Ms Mohini Mullick carefully honed her brief to match her requirements—100 m² in two levels of 50 m² each, set into the hill. On a steep south slope, the entry to the house is from the north at the first floor level where the living room and kitchen are located, and one climbs down for the bedrooms that are thus tucked into the hills. The structure is a load-bearing construction with a timber-framed roof.

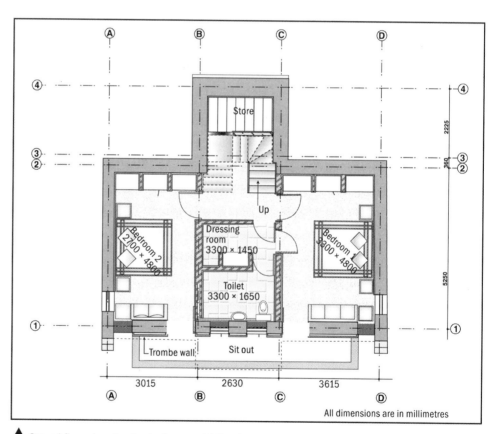

▲ Ground floor plan showing bedrooms with Trombe wall on the south

Features of the cottage

The cottage is oriented due south to maximize solar gain, and its compact shape reduces heat loss. The lower floor is provided with an earth cover, as the cottage is set partly into the hill. Direct solar gain is ensured in living/dining round and kitchen, which are the day-use spaces, by large south-facing glazed area, including inclined south-glazing. In addition to south-windows, a Trombe-like wall has been included on the south wall of the lower floor for indirect solar gain for bedrooms, being night-use spaces. The cottage is zoned so that all day-use spaces are on the upper floor and get more direct solar gain at the cost of some night losses.

▲ A view of the house showing south-facing glazed surfaces for the day-use spaces

▲ A view from inside the living spaces showing glazing for direct solar gains

The walls of the house are thick random rubble made from rubble available near the site. The joints are in cement mortar but kept very lean so as to give the look of dry rubble masonry. On the south side, this masonry provides a dark-coloured mass to store daytime heat, which slightly warms the bedrooms behind the south wall in the night. On the north, the house is sunk into the hill by an entire floor and this earth cover provides stable temperatures. On the east and west, the wall is left uninsulated. Insulating it from the inside would have drastically reduced the thermal mass of the building and insulating from the outside was aesthetically unappealing.

The entire roof is insulated with rock wool and the house is entered from the north through an air lock. There are a few openings on the east and west sides, and none on the north side of the cottage. The buffer spaces (lobby, stairs, etc.) are on the north.

Renewable energy systems

There is a 100 litre-per-day roof integrated thermosiphonic solar hot water collector system with the tanks located in the attic spaces. A wall-integrated counter-top operated solar food warmer/cooker is provided for the kitchen.

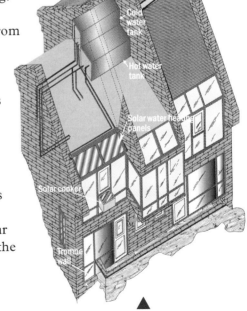

Axonometric view of the house showing south glazing and integrated solar water heater on roof and wall-integrated solar food warmer

User feedback The owner is satisfied with the thermal comfort provided by the cottage. The problems were largely due to the quality of construction, such as indifferent plumbing, leaking external walls, or shrinkage in timber, which needed attention over a long span of time. These problems persist in construction in the hills, especially due to lack of skilled man power.

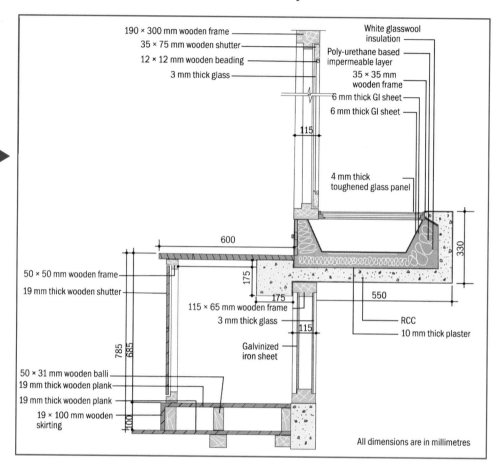

Sectional detail showing integrated solar food cooker and Trombe-like wall for indirect gain in bedrooms

At a glance

Project details

Project description Post-retirement cottage
Location Bhowali, Nainital, Uttar Pradesh
Architect Sanjay Prakash
Construction type Load-bearing structure (stone), with timber roof
Project period 1991–1995
Size 100 square metre covered area in a plot of about 200 m²
Client/owner Mohini Mullick
Builder/contractor Local contractor(s)

Design features

- Entry from north to give maximum solar exposure on south
- Entry through air lock
- Earth berming for lower floor by setting the cottage partly into hill
- Direct solar gain for living/dining room and kitchen by large south-facing glazed areas
- Indirect solar gain for night-use spaces
- Zoning to maximize solar gain for day-use spaces at the cost of some night losses
- Minimum openings on the east and west and no openings on the north.

Renewable energy systems

- Building-integrated solar hot water system and solar food cooker/warmer

MLA Hostel, Shimla

Editor's remarks

These hostel buildings have been designed to take maximum advantage of the sun. Innovative use of solar passive and active measures virtually negates the need for additional space-conditioning system. Building-integrated south-facing food warmers form a unique feature of the building.

Architects Himachal Pradesh State Public Works Department, Siddhartha Wig and Sanjay Prakash

Solar passive and active heating systems for a hostel building complex in Shimla

The MLA (Member of Legislative Assembly) Hostel is situated in close proximity to the Himachal Pradesh Vidhan Sabha building in Shimla, the state capital of Himachal Pradesh. Located in the cold and cloudy climatic zone, the design has to primarily cater to the difficult winter months. Heating and daylighting have been mainly considered while designing. Use of certain energy-efficient and renewable energy devices has also been suggested to increase the overall efficiency of the building.

The hostel is supposed to provide residential accommodation to the MLAs during assembly sessions. The whole complex comprises four blocks. The blocks are regular RCC-framed (reinforced cement concrete-framed) structures with brick in-fill walls. The blocks are oriented south (±15 degrees).

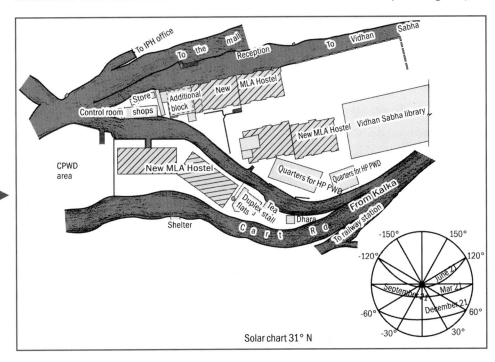

Site plan showing solar orientation of the MLA hostel buildings

At the time when Sanjay Prakash and Associates were appointed as solar architects for the project, work on Block 1 had already commenced and so the changes suggested were of retrofitting nature only. Since the design for Block 2 had also been done, the changes for that did not include major design and planning changes. The suggestions given helped evolve a more energy-conscious approach towards the design of blocks 3 and 4, rather than a retrofitting approach as had happened in the case of the earlier blocks. The state PWD (Public Works Department) has imbibed some of the passive solar design values, and this is apparent from the designs of blocks 3 and 4.

Categorization of concepts

All the proposed concepts have been categorized into three.

1. *Concepts for bringing up the specification from a low level to the normal level*   These were considered a standard part of normal building and these changes were not subjected to a cost-benefit analysis. (Examples of such concepts include providing a 9-inch thick brick wall instead of 4.5-inch wall for an external wall.)
2. *Concepts for improving the thermal performance of the buildings*   These formed the bulk of the passive solar suggestions and required a thorough cost-benefit analysis to gauge them for their efficacy. At times, these related to furnishing work, but mostly these related to civil work or services.
3. *Concepts to enhance the lifestyle of the inhabitants*   Most of these were in the form of appliances and required a cost-benefit analysis. These suggestions are mostly environmentally sound options.

Energy-saving features

An attempt has been made to lower the amount of conventional energy being used in running the buildings. Since cooling is not really a requirement, most of the suggestions are for heating or daylighting.

Fundamentally, the concepts kept in mind were

- orientation and compact planning
- solar gain and heat storage
- insulation, mass, and colour
- heat distribution and air movement
- use of energy-efficient and renewable energy devices.

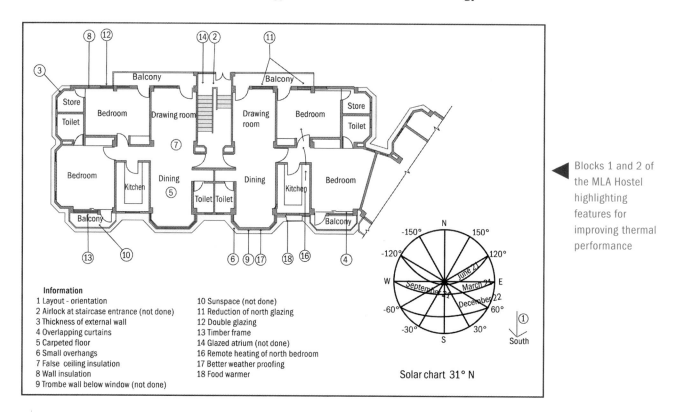

Blocks 1 and 2 of the MLA Hostel highlighting features for improving thermal performance

Information
1 Layout - orientation
2 Airlock at staircase entrance (not done)
3 Thickness of external wall
4 Overlapping curtains
5 Carpeted floor
6 Small overhangs
7 False ceiling insulation
8 Wall insulation
9 Trombe wall below window (not done)
10 Sunspace (not done)
11 Reduction of north glazing
12 Double glazing
13 Timber frame
14 Glazed atrium (not done)
16 Remote heating of north bedroom
17 Better weather proofing
18 Food warmer

Solar chart 31° N

Passive solar details

For this kind of climate, the design has to cater to the difficult winter months. Keeping this in mind a passive heating strategy has been devised. Use of certain energy-efficient and renewable energy devices has also been suggested because they increase the overall efficiency of the building and add to its character as an energy-conscious building.

The following is the list of suggestions, not all of which have been implemented, but has helped in improving the thermal efficiency of the building.

- *Revised layout* The buildings are oriented due south ±15 degrees for direct solar gain. They are spaced apart so as to eliminate shadows of one building falling over the other, even for the longer winter shadows. It was proposed that all bedrooms be south-facing to avail of the benefit of south exposure.

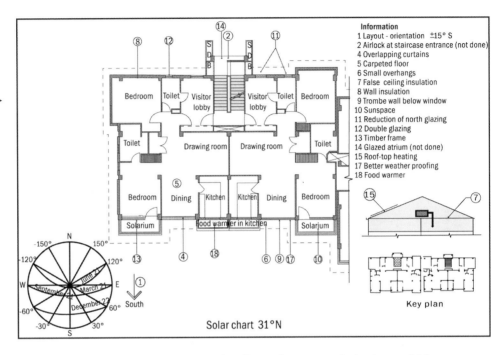

Blocks 3 and 4 of the proposed MLA Hostel highlighting features for improving thermal performance

- *Separate air-lock after staircase landing* Separate air-lock would have prevented heat loss, and decreased the rate of infiltration. This suggestion could not be followed for lack of space at the entrance.
- *Thickening of external wall* The existing external wall was changed to at least 9-inch brick wall or 12-inch stone wall as the existing 4.5-inch brick wall (Block 1) was inadequate for weather proofing.
- *Overlapping curtains* Well-sealed heavy curtains were used to act as thermal mass and to prevent heat loss.
- *Carpeted floors* Carpeted floor provide insulation and improve the general level of comfort. They should preferably be of dark colours when adjacent to south windows.
- *Small overhangs* Small overhangs helped to increase the amount of sunshine entering the building, while ensuring that no summer overheating took place. The primary purpose of the shades was rain protection. A 23-cm overhang can adequately protect a 1.2-m high south-facing window in peak summer while providing adequate rain protection.
- *Roof insulation* Roof insulation helped to preserve temperatures inside the building and prevented heat loss from the top floor. Rockwool insulation was provided above false ceiling.
- *Wall insulation* Wall insulation could be either insulation or a cavity wall and would help as in roof insulation. Eventually, the rockwool blanket/thermocole sheet was used in the walls behind panelling. Insulation was suggested on all walls except south because it was found out that the north, east, and west walls are net losers of heat.
- *Sunspace* The existing balcony can be made into a sunspace resulting in increased heating especially during the winter months. This suggestion could not be incorporated in blocks 1 and 2 because of prior structural limitations, but has been introduced in blocks 3 and 4.

- *Reduction of north glazing* Since solar heat gain through north-facing windows is negligible, glazing on the north increases heat loss to the ambient. This is rectified by reducing the amount of glazing in the north, and providing double glazing in the essential windows.
- *Double glazing* There is the inherent problem of condensation in badly executed double glazing, which can be seen in some earlier buildings in Shimla. This is overcome by providing two separate sets of shutters with glazing that allow cleaning. Alternatively, hermetically-sealed double-glazing with a desiccant strip or vacuum-sealed windows can be used but they are expensive.
- *Plastic/timber frames* Since plastic and timber have a lower conductivity than steel it is advisable to use them for joinery to reduce conductive heat loss to the outside. Steel would aid heat transfer and would work against attempts to retain heat inside the building.
- *Glazed atrium over staircase* Since the staircase is on the north side this would have helped to improve the general level of lighting in the staircase well. This could not be incorporated because of lack of space on the ground adjacent to the staircase (blocks 1 and 2).
- *Remote heating of north bedroom* This idea was proposed to tap heat on the south wall and transport it to the north bedroom. Small Trombe walls below and on the sides of the kitchen window trap the heat, which is conveyed through a duct to the north bedroom. A small fan can be used to assist air movement.
- *Better weather proofing* These measures at the openings helped to reduce infiltration. Rubber stripping at the junctions of shutters and frames and below doors helped in sealing of all cracks. Keeping windows closed, and immediately replacing broken panes especially during winter months would help to considerably reduce infiltration.
- *Food warmer in kitchen* This is a feature to use solar heat to keep food in the kitchen warm. Food warmer has been provided outside the window of kitchen towards southern side.

Renewable energy system

Solar water heaters
These can be mounted on the roof and used to tap the sun to heat water. The solar water heaters are now in the process of installation.

Low wattage electrical radiating panels

These can be used to augment the available heat during peak winter. Also they would form an integral part of the heating strategy in the apartments that do not receive any sunlight (Block 4) where some mutual shading takes place. This is not a building-related suggestion and can be done by the inhabitants themselves.

Performance results Performance monitoring has been carried out by Technical Cell; State Council for Science and Technology and Environment, Himachal pradesh in this building. In February 2000, the inside temperatures range between 10 °C and 26 °C corresponding to ambient temperatures between 4.5 °C and 7 °C. The food warmer temperature goes up to 32 °C during this period.

Users' feedback
- The food warmer is a useful device. The residents use it regularly, hence saving the conventional energy (electricity or gas) to warm food.
- Electrical convectors are required in winter, but their use starts much later (from October or November instead of September).
- The provision of a small office/visitors' lobby is very much appreciated.
- The view outside is restricted, which is a major drawback in a place like Shimla.
- There is no place for drying clothes in the apartment, nor any common space provided for this in the hostel.

 South side of Block 3 of MLA Hostel highlighting the solar food warmers integrated with the building

At a glance

Project details

Name of the project MLA Hostel, Shimla
Climate Cold and cloudy
Promoters Technical Project Management Cell, State Council for Science, Technology and Environment, Shimla
Year of completion 1999
Client/owner Government of Himachal Pradesh
Cost The initial cost of Rs 10 723/m² increased to Rs 11 178/m² due to incorporation of passive solar measures. Thus, there was a 4.2% increase in overall costs over the initial estimates

Design team
Solar architects Sanjay Prakash and Siddhartha Wig
Local architect B P Malhotra, Chief Architect, Himachal Pradesh Public Works Department; Surinder Kumar, Senior Architect, Himachal Pradesh Public Works Department

Design features (proposed although some were not executed due to various constraints mentioned elaborately under each head)

- Solar orientation
- Air-lock after staircase landing to prevent heat loss and infiltration
- Enhanced thickness of external wall
- Heavy curtains and carpeted floors to add to thermal comfort
- Adequately sized overhangs to maximize solar access in winter
- Roof and wall insulation
- Trombe wall and sunspaces at appropriate locations
- Appropriately sized and detailed glazing system
- Glazed atrium over staircase
- Innovative heating systems using solar water and air heaters
- Better weather stripping to reduce infiltration
- Integrated food warmer with the structure
- Roof mounted solar water heater
- Low wattage electric radiating panels for back-up heating

Temperature check for Shimla

(see Appendix IV)

Cold	Cool	Comfortable	Warm	Hot	
	•				January
	•				February
	•				March
	•				April
		•			May
		•			June
		•			July
		•			August
	•				September
	•				October
	•				November
	•				December

Add the checks

	8	4		

Multiply by 8.33% for % of year

Heating	67
Comfortable	33

Climatic zone: cold and sunny

Degree College and Hill Council Complex, Leh

Airport and staff housing colony, Kargil

LEDeG Trainees' Hostel, Leh

Sarai for Tabo Gompa, Spiti

Degree College and Hill Council Complex, Leh

Editor's remarks

Designed for the cold desert of Leh, these buildings promise to give thermal and visual comfort by judicious use of solar energy.

Architects Arvind Krishan and Kunal Jain

Sun to give warmth to the occupants of these buildings in the cold desert of Leh

Located at Leh at an altitude of 3514 m above mean sea level in the 'upper' Himalayas, the Degree College and the Hill Council Complex have been designed and built in a cold and sunny climate zone experiencing a large number of sunny days.

A view of the lecture theatre in the academic block with a completely solid north block and south side designed to maximize solar penetration for heating and daylight distribution

Lying in the cold and sunny climate zone, Leh has a fairly long winter—October to March, almost well into April. The severe cold spell lasts about two months (minimum dry bulb temperature –30 °C) and there is very little precipitation throughout the year. June and July are quite pleasant. Although the region can be described as dry, recently there has been recorded incidence of increasing precipitation. The region has good regime of sunny days with high radiation levels.

Degree college

The site The site for the Degree College is an elongated rectangle with the land sloping towards south at a gradient of 1 in 30. The site is characteristic of the cold dry desert condition of Ladakh—by and large barren with no vegetation. It gives an excellent view of the snow-capped mountains that surround the site on all sides.

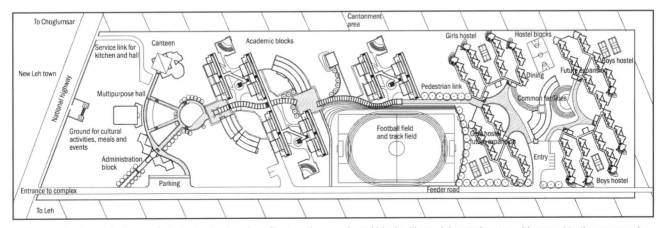

A site plan of the Degree College showing location of lecture theatres, hostel blocks, library, laboratories, etc. with respect to the campus plan

Passive solar features

Lecture theatres, laboratories, and library in the academic block have been designed with the building section optimized for both heat and daylight penetration. The north side of the complex has been designed with solid walls in each building to eliminate heat loss, while the south side has been designed to maximize solar penetration for heating and daylight distribution.

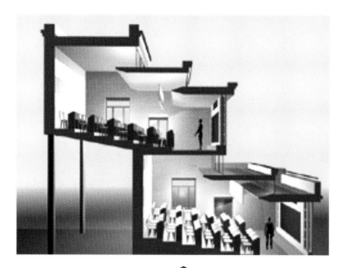

A section of the lecture theatre showing solar access and daylighting design

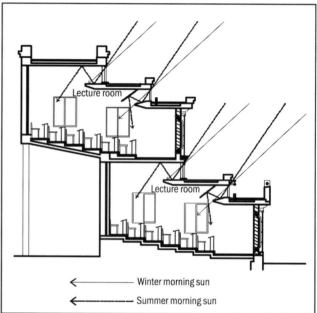

At a glance

Project details

Site Elongated rectangle with the lie of the land sloping towards south at a gradient of 1 in 30
Building type Institutional
Climatic zone Cold and sunny
Status Ongoing (Final project to comprise academic and administration block for approximately 1000 students, hostel complex for 400 students, library, multipurpose hall, sports track, canteen, etc., with a total built-up area of 63 100 m² in a site area of 9030 m²).
Built-up area 8250 m²
Completion 1998 (Phase I)
Contractor Local

Design features

- Building section optimized for both heat and daylight penetration
- North side designed as solid walls to reduce heat loss
- South side designed to maximize solar penetration for heating and daylighting

Hill Council Complex

The site

The site for the Hill Council Complex lies on a southernly sloping land with an average gradient of 1 in 12. High snow-capped mountains surround the site on all sides, with the Leh Palace forming the backdrop.

A computer-generated view of the Hill Council assembly building at Leh

Passive solar features

Passive solar heating

Both plan and three-dimensional form of the buildings allow maximum penetration of sun, that is both solar heat gain and daylight are maximized. While heat gain is maximized, its absorption in the judiciously distributed thermal mass provides heat in the spaces during the diurnal cycle. Air-heating panels, designed as an integral part of the southern wall, provide effective heat gain through a close connective loop.

Features to minimize heat losses

Good insulation and minimum fenestration on northern exposure prevent heat loss. Infiltration losses are minimized through weather-proofed (i.e. with no thermal bridges) wooden joinery. Double-glazing helps control heat loss from glazing without creating condensation.

A computer-generated section through the assembly hall of the Hill Council assembly building

Ventilation

Since solar heat gain may raise the internal ambient temperature (as per simulation prediction) above comfort range in summers, ventilation is an effective strategy for summers to dissipate internal heat build-up. To optimize ventilation, the convective loop is coupled with cross-ventilation through eastern/western fenestration. Consequently, the institutional building requires no electrical energy for heating, lighting in daytime, and for hot water since this can be achieved through a solar water heating system.

Ventilation is achieved through the convective loop activated by buoyancy. This building can, therefore, be described as a 'zero energy consuming' building during daytime, the normal hours of operation.

At a glance

Project details

Building type Institutional
Climatic zone Cold and sunny
Location Leh, Ladakh
Area 2400 m²
Completion 1998 (Phase I) Final project to comprise assembly hall, offices for the main administrative body of the Hill Council of Ladakh and conference facilities with total site area and built-up area of 29 600 m² each.
Contractor Jammu and Kashmir PCC

Design features

- Plan and three-dimensional form of the building maximize solar heat gain and daylight
- Air-heating panels provide effective heat gain through a close connective loop
- Good insulation and maximum fenestration on northern exposure
- Infiltration losses minimized through wooden joinery
- Cross-ventilation during summer to avoid heat build-up
- Solar water heating system
- Zero energy consumption during normal hours of operation

Temperature check for Leh

(see Appendix IV)

Cold	Cool	Comfortable	Warm	Hot	
•					January
•					February
•					March
	•				April
	•				May
	•				June
		•			July
	•				August
	•				September
	•				October
•					November
•					December

Add the checks

5	6	1		

Multiply by 8.33% for % of year

Heating	92
Comfortable	8

Airport and staff housing colony, Kargil*

Architects Arvind Krishan and Kunal Jain

An airport building and staff housing colony designed for maximum solar advantage in the cold sunny climate of Leh

Airport terminal complex

The Kargil Airport Terminal Complex has been designed to meet the requirements for a smaller airfield within the remote region of Kargil. Facilities provided for are the arrival and departure lounge, waiting hall, office, and VIP lounge and ancillaries.

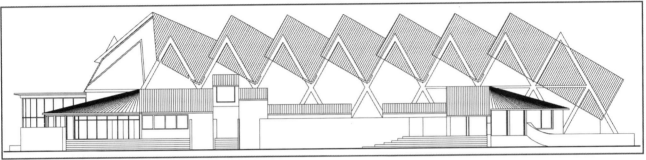

▲ A computer generated form of the airport at Kargil. The form has been developed to give maximum solar access

Passive solar features

Direct and indirect solar gain

The building has been designed for maximum solar penetration in all spaces during the critical periods with the major spaces being heated with solar energy by use of both direct gain as well as indirect gain. Further, the building has been optimized for uniform and glare-free light distribution in the office areas.

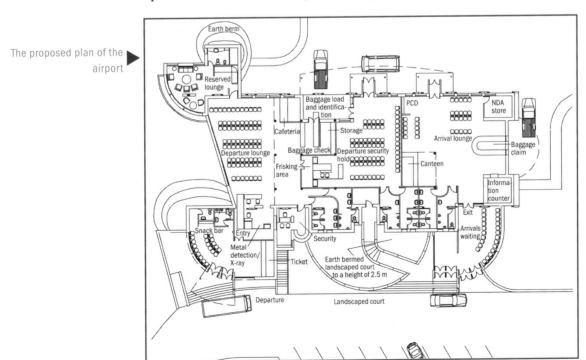

The proposed plan of the airport ▶

---

*As this project is currently under way, no editorial comments are provided

Insulation

Maximum solar exposure is achieved by large double-glazed surfaces on the south face. The 500-mm thick composite walls (300 mm stone outer veneer + 50 mm insulation + 150 mm cement concrete hollow block on the inside veneer) on the other three directions increase the thermal lag and insulate the interior from the harsh outdoor environment. This section has proven very effective for thermal performance in this climate.

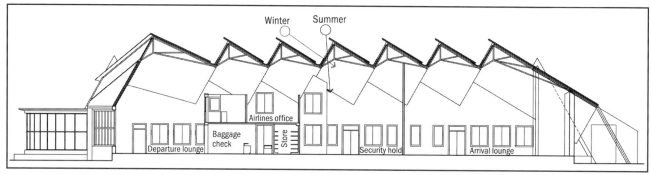

▲ A section of the airport showing change in solar access due to specially designed roof openings. The spaces are all designed to be daylit thus reducing the need for artificial lighting

Earth berming

The waiting halls and earth berming on the east face further act as insulation buffers.

Roof details

The clear storey glazing at roof level on the south maximizes solar heat gain and daylight to all inner spaces with adequate penetration of sun. The roof angle has been designed to prevent mutual shading during the critical periods. Adequate eye level light and view is achieved by the west façade (airfield side).

Materials

The building is an innovative articulation of local materials (stone, mud, wood) and elements (brackets, motifs) with modern materials (reinforced cement concrete, steel, glass, etc.) and seismic-resistant structural systems (steel A-frame support system with an intricate steel space truss roof).

At a glance

Design features

- Double-glazed surfaces for the south face
- 500-mm thick composite walls on the other three sides increase the thermal lag.
- Earth berming on the east face provides further insulation
- Clerestory maximizes solar heat gain and daylight to all inner spaces
- Innovative mix of materials

Project details

Building type Airport
Location Kargil, Ladakh
Architects Arvind Krishan and Kunal Jain
Climate zone Cold and sunny
Contractor Local
Status Ongoing
Built-up area 1320 m²

Airport staff housing, Kargil

The airport staff housing provides for 14 individual residential units: 2 three-bedroom units, 4 two-bedroom units, and 8 one-bedroom units. Designed in a cluster-like compact form that maintains the individuality and privacy of each unit, maximum solar penetration has been provided in all main spaces during the critical periods.

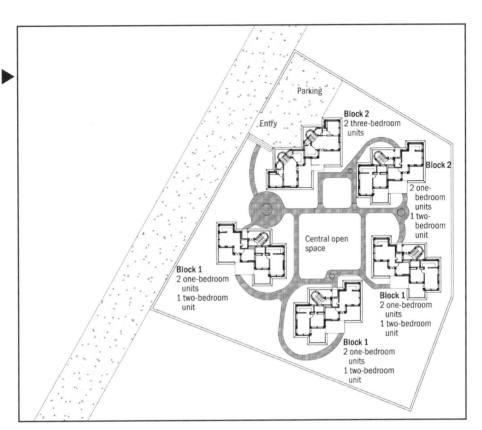

Building plan showing a cluster of houses in the staff housing colony

Orientation

Clusters have been designed around a central open space with maximum southern exposure. Terraces and glazing on the longer side have southern orientation for direct solar gain and minimum openings on the north-wall prevent heat loss.

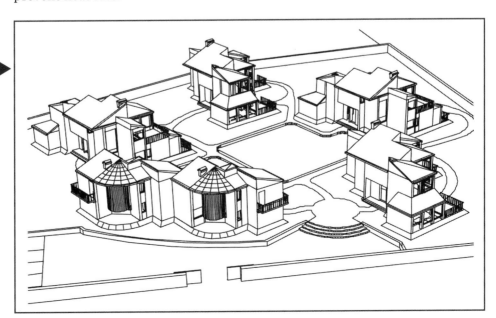

The proposed view of staff housing colony

Airlock

All north-side entrances have been provided with a double airlock and all the service spaces – kitchen, toilets, staircases – are placed on the north while the living spaces face south.

Insulation

The composite wall section of concrete blocks, insulation layers and stone has been used to increase thermal lag.

At a glance

Design features

- Clustered around central open spaces with maximum southern exposure.
- Terraces and glazing on the longer side have southern orientation for direct solar gain
- Minimum openings on the north wall prevent heat loss.
- North-side entrances provided with double air lock.
- Service spaces are placed on the north and the living spaces face south.
- Composite wall section of concrete blocks, insulation layers, and stone used to increase thermal lag.

Project details

Building type Residential
Location Kargil, Ladakh
Architects Arvind Krishan and Kunal Jain
Climate zone Cold and sunny
Status Proposed
Site area 4492 m²
Built-up area 1003 m²
(proposed)

Temperature check for Kargil

(see Appendix IV)

Cold	Cool	Comfortable	Warm	Hot	
●					January
●					February
●					March
	●				April
	●				May
	●				June
		●			July
	●				August
	●				September
	●				October
●					November
●					December

Add the checks

5	6	1		

Multiply by 8.33% for % of year

Heating	92
Comfortable	8

LEDeG Trainees' Hostel, Leh

Architect Sanjay Prakash

Solar passive hostel in cold and sunny climate with a variety of indirect heat gain systems

Editor's remarks

A training hostel for students and trainees, the building uses various types of direct and indirect heat gain systems. The performance of each system has been monitored and the results help the designer to choose the right kind of system for new constructions.

LEDeG (Ladakh Ecological Development Group) is a well-known non-government organization operating out of various centres in Ladakh. This project is located in LEDeG's Chans'pa Centre, in Leh.

Leh's characteristically cold and sunny climate (temperatures below –30 °C and has over 320 sunny days in a year) makes it a testing ground for solar technologies despite its small population. It is also a region where harnessing solar energy can lead to great human welfare at very little cost.

▲ The south façade with Trombe walls caters to heating needs of the rooms

The LEDeG Hostel provides sleeping accommodation for 24 persons, with toilet, laundry, and study facilities all integrated into the building. The hostel is two-storey with 12 double bedrooms and ancillary spaces. It is set in the northern part of the Chans'pa campus, on a slightly south-facing slope.

Traditional techniques have been modified and adapted for use in the building. The load-bearing walls of the ground floor are made in rubble masonry with mud mortar. The upper floor uses load-bearing sun-dried mud bricks (adobe) in mud mortar. Partition walls are also executed in adobe. The intermediate floors and roofs are timber-framed, with poplar wood joists covered by twigs, grass, and earth. In order to combat a rising local fear of increasing leakage in such construction, a small slope has been given to the flat roof so as not to create a sag in the middle.

Materials, techniques, and methods

South orientation

The building was oriented south in the sense that every habitable room has a liberal south exposure. Only corridors, toilets, and staircases are without direct south orientation. The 12 bedrooms and the laundry and study rooms all have large south exposure for winter heat gain with almost no overhang. Almost all winter heating needs can be met by the south exposure. There is a slight play with the exact south orientation, in that the two wings of rooms are slightly angled to each other, thereby demonstrating that solar orientation need not be interpreted in a strict orthogonal grid layout, and can allow some latitude for variations in orientation.

Plan showing solar orientation of building ▼

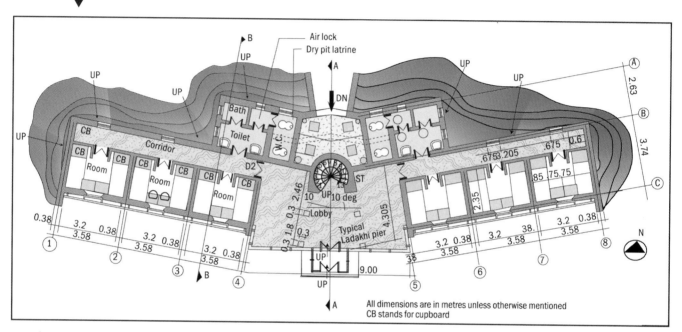

Solarium

At the centre of the building is a south-approached air-locked entrance leading to the study room. Above this is another study room. Either of these can be occasionally used as a dining space. Since these spaces are used at daytime only, they are heated by south-glazing tilted at 60 degrees to the horizontal. The directly gained solar heat is stored in the mass of the building and the warmth can remain for a few hours after sundown.

Trombe wall

The bedrooms have been provided with a mix of windows and glazed walls. Some of these are the classical Trombe wall type design with vents for convective loop formation. Others are provided without vents so as to compare their behaviour, especially as heat is not required in the bedrooms until early in the evening. These mass walls are made of rubble (ground floor) and adobe (first floor).

Insulation and mass

The thick earth in walls and roof provides both insulation and mass.

Building section showing integration of renewable energy systems into the building. Also seen is the Trombe wall and solarium

Absorbing finish

The south walls are painted black to provide better heating. This colour, along with maroon, has been used to embellish all openings. The local traditions have been respected while providing for better energy efficiency.

Weather-stripping

All openings have cork-based weather-strips at the edges to seal them tight.

Solar hot water

A flat-plate thermosiphonic collector system is provided on the roof of the building. At an angle of 60-degree tilt, it provides year-round hot water.

Composting pit latrine

Two sets of sloping pits with vent pipes, attached to the building, provide a dry latrine at the first-floor level. Cleaning is possible at ground-floor level. This feature is important in that it can reduce the incidence of frostbite that is common in Ladakh, partly because the people need to go out to the field or an external pit latrine for their morning ablutions.

Daylight

All spaces in the building are properly lighted with natural light in the day.

Future provision of photovoltaic charging

It is planned that the building shall be fitted with photovoltaic cells to provide most of the night lighting.

Performance The thermal systems worked extremely well in winter, eliminating the requirement of double quilts altogether. The temperatures inside the sleeping rooms stayed above 8 °C during a moderate winter when outside temperatures dropped to –17 °C.

Hot water systems have worked well. Pit latrines have not worked well. The vent pipe is probably not large enough to eliminate the smell in the thawing season of spring. The pits being placed on the north of the building also remain cold, adding to the problem. Other variations in design are being tried out.

The hostel at LEDeG's centre in Chans'pa ('Shanti Stupa' Centre) is used for students and LEDeG employees who reside outside of Leh, but are staying in town for training or work. It has 12 rooms and uses four different passive solar heating strategies. These include two half-Trombe walls, four unvented Trombe walls, four full vented Trombe (no backflow prevention) walls, and two direct gain (glass) rooms.

The performance results of the rooms for the month of January 1997 and for a week in July 1997 are summarized.

Summary of solar thermal evaluations*

All of the rooms remained above 0 °C at all times in January, even when the temperature outside fell to –17 °C. However, the winter of 1997 was particularly warm, and performance will differ in colder years.

A summary of the performance of the four rooms in January 1997 is given in Table 1. These data show that the unvented and vented Trombe walls had very similar performance. They both stayed quite warm throughout the month, and the temperature in the rooms was stable. The unvented Trombe was more stable than the vented Trombe, and it also has the advantage of less maintenance, as less dust collects on the blackened Trombe surface due to the lack of vents.

The half-Trombe was more stable than the direct gain room, and it stayed slightly warmer at night. However, the improvement in performance is only marginal. The main benefit may be that it does not get quite as warm during the daytime, due to increased thermal mass, while still allowing a significant amount of light into the room.

The reason for the low temperature during the night is probably due to heat loss through the top of the window (the half-Trombe wall is on the bottom, with the window going up to the ceiling at the top). The half-Trombe and the glass room also have more heat loss than the full Trombe walls, as they have three exterior walls, instead of just one.

Table 1 Passive solar heating performance data for January 1997

Room	Average temperature (°C)	Minimum temperature (°C)	Maximum temperature (°C)
Half Trombe	7.9	2.8	17.0
Unvented Trombe	13.2	9.1	16.9
Vented Trombe	13.2	8.1	18.8
Direct gain	7.0	0.3	23.3
Outside air	–8.3	–16.6	1.2

Summer performance of the passive heating technologies is also important, especially with regard to overheating during the day. The performance of the rooms for a week in July is given in Table 2. These data show that all of the rooms are quite warm in summer. Again, the full Trombe walls are the most stable, with greater variations for the half Trombe and direct gain rooms. The rooms had people staying in them during the January and July data collection periods presented here. The lower minimum temperature for the vented Trombe (19.1 °C) in July is probably due to the occupants leaving the windows open to keep cool.

---

*The source of the performance monitoring results is *Renewable Energy Resource Data Collection in Ladakh, India* (A summary of data collected and other work completed in 1996), Ladakh Ecological Development Group, Leh, Ladakh, August 1997.

Table 2 Passive solar heating performance data for one week in July 1997

Room	Average temperature (°C)	Minimum temperature (°C)	Maximum temperature (°C)
Half Trombe	23.6	20.4	26.1
Unvented Trombe	24.4	21.3	26.4
Vented Trombe	23.8	19.1	26.7
Direct gain	25.8	22.1	29.0
Outside air	21.6	12.0	31.2

A decision about which of these room designs is the most appropriate is a subjective one depending on the preferences of the user. If the summer temperatures for the rooms seem too warm, then cooling strategies such as overhangs that shade the glazing (glass) in summer, but not in winter, should be used. Passive ventilation strategies can also be useful to prevent overheating in summer.

For winter performance, again the decision must be made according to the tastes of the user. The unvented Trombe presented here heats the room successfully, maintaining a stable and warm temperature throughout winter without any additional heating. The other rooms also provide passive heating to different degrees. If a stable temperature is the most important criteria, then an unvented Trombe wall is recommended. Direct gain rooms are probably more appropriate for spaces that are for daytime use only (such as offices), while rooms that are to be kept warm throughout the night are best served with a Trombe design. In addition to residential bedrooms, the Trombe design should be considered for power houses for diesel and hydro-electric equipment, battery rooms, and other places where equipment should be kept warm.

Economics

In Ladakh this type of architecture could be called modern accepted practice, hence its cost was lower than the normal construction due to better management and use of local construction techniques. The payback period of even the expensive components like glass is as low as one winter season. The building uses a negligible amount of artificial heating by bukharis today, whereas it would have typically consumed over 6 tonnes of firewood to heat every winter.

At a glance

Project details

Project description Hostel building for trainees in appropriate technology
Architect Sanjay Prakash
Climate Cold and sunny
Consultants In-house
Project period 1994–1996
Size 300 m² covered area in a small campus
Client/Owner LEDeG (Ladakh Ecological Development Group)
Builder/Contractor Owner-managed construction

Design features

- Traditional materials and methods of construction have been modified and adapted to achieve energy efficiency
- Predominantly south exposure with no overhangs for maximum winter gains.
- Entrance lobby designed as a solarium on the south side.
- Bedrooms provided with various types of Trombe walls (half Trombe, unvented Trombe, vented Trombe) or direct gain systems for passive heating.

Temperature check for Leh

(see Appendix IV)

Cold	Cool	Comfortable	Warm	Hot	
●					January
●					February
●					March
	●				April
	●				May
	●				June
		●			July
	●				August
	●				September
	●				October
●					November
●					December

Add the checks

5	6	1		

Multiply by 8.33% for % of year

Heating	92
Comfortable	8

Sarai for Tabo Gompa, Spiti

Editor's remarks

Located in the cold desert climate of Spiti, this building focuses on use of local materials and methods of construction. Incorporation of various passive features provides thermal and visual comfort to the residents.

Architect Shubhendu Kaushik

Located in a climatic zone where the temperature ranges between 35 °C and −25 °C, this small building for the Tabo Gompa is a low-energy, low-cost building using various passive techniques for winter heating

The skylight of Tabo Sarai for daylight integration and solar gain. The inset shows the view of the Trombe wall

Passive solar features

Trombe wall, light well/sunspace and thermally massive construction

The Sarai, located in the Spiti sub-division of Lahaul and Spiti district, is an institutional building for the Tabo Gompa. Due to site constraints, the building had to be oriented to face east and west. The south wall has been designed as a Trombe wall. However, in order to make the Trombe wall aesthetically pleasing, the frame has been painted in vibrant colours.

Two intermittent courtyards, which are covered by glass, act as sunspace as well as light wells. These sunspaces provide heat and light for the northern depth of the building.

The north-east and west walls are thermally massive construction made of 2-feet thick rammed earth. The west wall, which is on the windward side, has

A cross-sectional view through the central courtyards meant for heat gain and daylighting ▼

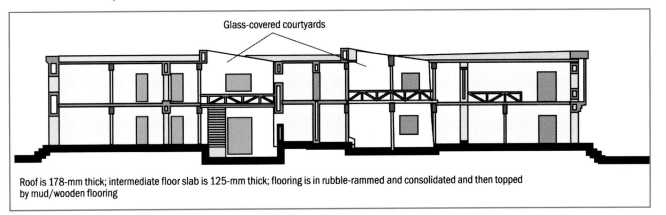

Roof is 178-mm thick; intermediate floor slab is 125-mm thick; flooring is in rubble-rammed and consolidated and then topped by mud/wooden flooring

A schematic view of the building plan showing arrangement of rooms around central courtyards

minimum possible openings to protect from the cold winds. The north and east walls also have minimum openings and most openings are towards the sunspace courtyards or in the Trombe wall. The openings in the walls have been planned as double glazing with wooden shutters opening inwards.

Materials and methods of construction

Foundations up to the plinth level have been made of random rubble stone masonry in mud mortar while rammed earth is used for the walls. The roofing is traditional Spiti mud-roof on wooden framework. The flooring is done with either mud or wood. Mud plaster with an external coating of lime wash has been used. Local water resistant mud 'Tooah' has been used on exterior walls and exposed roof.

South wall of the Sarai for Tabo Gompa highlighting the aesthetically designed Trombe wall

At a glance

Project details

Building/project name Sarai for Tabo Gompa
Location Tabo village, Spiti sub-division of Lahaul and Spiti District
Climate Cold and sunny
Building/construction type Predominantly rammed earth construction
Building category Institutional
Architect Shubhendu Kaushik
Year of start 1994
Sponsor Tabo Gompa and Government of Himachal Pradesh

Design features

- Unfavourable orientation due to site constraints
- Trombe wall on the south
- Minimum openings on north, east, and west walls
- Thermally massive construction
- Centrally located courtyards covered by glass for heating of inner rooms and daylighting

Climatic zone: composite

Residence for Madhu and Anirudh, Panchkula

PEDA office complex, Chandigarh

Bidani House, Faridabad

Transport Corporation of India Ltd, Gurgaon

SOS Tibetan Children's Village, Rajpur, Dehradun

Redevelopment of property at Civil Lines, Delhi

Integrated Rural Energy Programme Training Centre, Delhi

Tapasya Block (Phase 1), Sri Aurobindo Ashram, New Delhi

Residence for Sudha and Atam Kumar, Delhi

Residence for Neelam and Ashok Saigal, Gurgaon

Dilwara Bagh, Country House for Reena and
Ravi Nath, Gurgaon

RETREAT: Resource Efficient TERI Retreat for Environmental
Awareness and Training, Gurgaon

Water and Land Management Institute, Bhopal

Baptist Church, Chandigarh

Solar Energy Centre, Gual Pahari, Gurgaon

National Media Centre Co-operative Housing Scheme, Gurgaon

American Institute of Indian Studies, Gurgaon

Residence for Madhu and Anirudh, Panchkula

Architects Anant Mann and Siddhartha Wig

A small residential house in the composite zone that uses simple economically viable solutions to respond to climatic needs

Editor's remarks

This small residential unit located in the composite zone uses simple, cost-effective interventions for achieving energy efficiency. These interventions aim to inspire other residences to innovate and adapt such techniques at the design stage itself so as to save on electricity bills in future.

The city of Panchkula lies in the plains at the foot of the lower Himalayas, within a composite climatic context.

The building is oriented due south. The insulated west wall clad in slate can be seen. The black solar chimney and light shelves on the south windows are also visible

Panchkula experiences wide climatic swings over the year, i.e. very hot and dry period for almost two-and-a-half months (maximum DBT [dry bulb temperature] 44 °C) and quite cold period for a shorter duration (minimum DBT 3 °C). The hot dry period is followed by a hot and humid monsoon

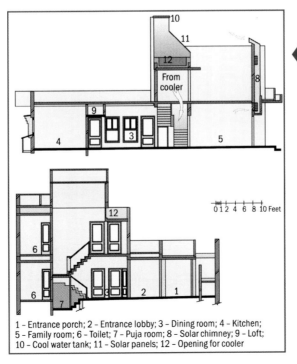

Section of the building showing centralized evaporative cooling system for the rooms. The air movement is assisted by the solar chimney

1 - Entrance porch; 2 - Entrance lobby; 3 - Dining room; 4 - Kitchen;
5 - Family room; 6 - Toilet; 7 - Puja room; 8 - Solar chimney; 9 - Loft;
10 - Cool water tank; 11 - Solar panels; 12 - Opening for cooler

Residence for Madhu and Anirudh, Panchkula

Building plan showing angular design of the house in order to give southern exposure to most of the rooms

1 Entrance porch
2 Entrance lobby
3 Living room
4 Family study room
5 Dining room
6 Kitchen
7 Bedroom
8 Front yard
9 Puja room
10 Toilet
11 Walk in closet
12 Court
13 Backyard

(A) Insulated west wall
(B) Solar chimney

period of about two months (maximum DBT 38 °C and maximum relative humidity 90%), with intervening periods of milder climate. The demand on building design, therefore, is to respond to the extremes: eliminate (minimize) heat gain in hot and dry period, maximize ventilation in hot and humid period from zones/areas designed as heat sinks, and maximize heat gain in cold period. This is a small residential building on a rectangular plot measuring 245 m² and opening on to the south-west. The rooms are placed around a courtyard with the master bedroom and the study/family room facing the south.

The living room faces north while the second bedroom on the ground floor opens on the north and gets direct sunlight from the court. The dining room is also adjacent to the courtyard. The structure is load-bearing brick with RCC slab. The woodwork in doors and windows is in sheesham wood and the floors are brick and polished stone.

The windows and door of the dining area face the court. Also the rear bedroom gets its southern exposure from the court

Passive solar features

The main building was oriented due south to have better control of the sun. Properly designed south shades coupled with light shelves determine the south facade. The courtyard ensures good lighting in the dining and the rear bedroom. The orientation ensured winter sun (while keeping the summer sun out) and adequate daylight in the rooms. The adjacent building keeps the east sun out while the west wall is insulated with 50 mm of polystyrene.

Light shelves and white ceilings in the rooms allow for enough daylighting in the rooms. There is a central evaporative cooler above the staircase well. This air current is augmented by a solar chimney which extracts air from the front room. Louvres in the doors ensure air circulation even if the doors are closed.

The sloping slab above the stairs allows for architectural integration of solar water heating panels or solar photovoltaic panels as the case may be. The north face of this well catches more light and carries it down the stairs.

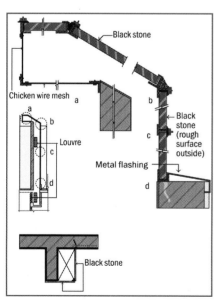

Section of the solar chimney for ventilation

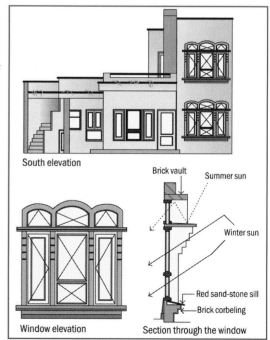

Windows allow winter sun and cut off harsh summer sun

At a glance

Project details

Project description Residence for Madhu and Anirudh located at Panchkula

Architects Anant Mann and Siddhartha Wig

Project period 1999

Owners Dr Madhu and Anirudh Khullar

Design features

- Orientation to catch winter sun and keep away summer sun
- Proper shading and daylighting
- Solar chimney for ventilation
- Insulation on west wall

Installed systems

- The evaporative cooler with a 24-inch fan size and 2.5-m² pad area would provide central evaporation cooling

Renewable energy systems

- Provision for solar hot water system
- Provision for solar photovoltaic system

Economics

- Additional costs of orienting the building have not been assessed since they are difficult to verify
- Additional costs of west wall insulation, solar chimney, and insulated pipes constitute less than two per cent of the building cost
- Since only the future provision for solar hot water and photovoltaic panels is provided, their cost has not been assessed

PEDA office complex, Chandigarh *

Architects Arvind Krishan and Kunal Jain

Solar architecture in an urban context with rigid architectural controls

The climate

The PEDA (Punjab Energy Development Agency) office complex is located in Chandigarh, on a practically square site that lies on flat land with no major topographical variations. Chandigarh experiences wide climatic swings over the year, i.e. very hot and dry period of almost two and a half months (maximum DBT [dry bulb temperature] 44 °C) and quite cold period of a shorter duration (minimum DBT 3 °C). The hot dry period is followed by a hot humid monsoon period of about two months (maximum DBT 38 °C and maximum relative humidity 90%), with intervening periods of milder climate.

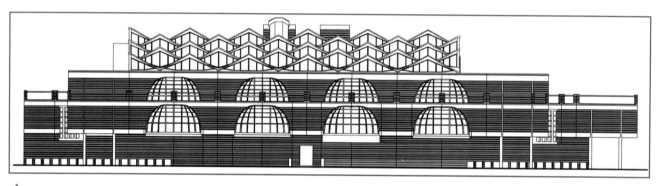

▲ South elevation showing domical roofs and vertical roof glazing systems for daylight integration and ventilation

Design response

The demand on building design, therefore, is to respond to the extremes: eliminate (minimize) heat gain in the hot-dry period, maximize ventilation in hot humid period from zones / areas designed as heat sinks and maximize heat gain in the cold period. Within the context of the radical experiment that is Chandigarh, the PEDA building has been designed with an ethos: design with nature. The physical context although unique in itself, i.e. the urbanity of Chandigarh, offers yet another challenge for design.

The site

The site is located on a major road intersection and lies on the edge of a residential area with other proposed office buildings on the other edge.

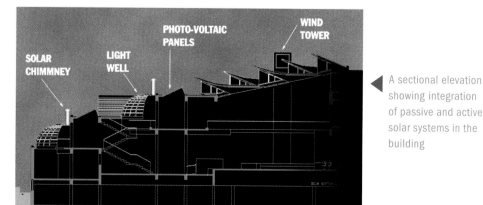

◄ A sectional elevation showing integration of passive and active solar systems in the building

---
* As this project is currently under way, no editorial comments are provided.

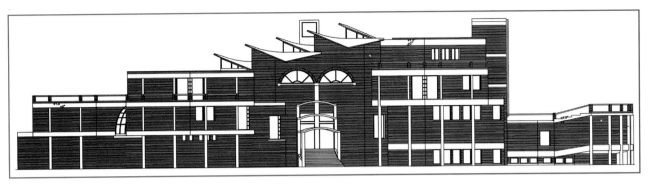

▲ Interconnected volumes of space to enable passive space conditioning of entire volume of building

Scale form

While the three-dimensional form of the building has been developed in response to solar geometry, i.e. minimizing solar heat gain in the hot-dry period and maximizing solar heat gain in the cold period, the scale and form of the building responds to its urban context as well. Whereas, expression of the building on the two main roads of the intersection bears the character and scale of an office building, the building responds to the residential context on the south/south-east edges by gradually scaling down in mass and volume.

Climate responsive building form

Light wells, solar chimneys, and wind towers

To achieve a climate-responsive building, an innovative concept in architectural design has been developed. In place of the 'central loaded corridor' plan stacked on top of each other to make various floors, which has become virtually the generic form for an office, the PEDA building is a series of overlapping floors at different levels in space floating in a large volume of air, with interpenetrating large vertical cut-outs. These vertical cut-outs are integrated with light wells and solar-activated naturally ventilating, domical structures. This system of floating slabs and the interpenetrating vertical cut-outs is then enclosed within the envelope of the building. The envelope attenuates the outside ambient conditions and the large volume of air is naturally conditioned by controlling solar access in response to the climatic swings, i.e. eliminating it during hot-dry period and maximizing its penetration in cold period. The large volume of air is cooled during the hot period by a wind tower, integrated into the building design, and in the cold period this

◀ A computer generated image showing sectional view of the domical roof for ventilation and daylighting

volume of air is heated by solar penetration through the roof glazing, generating a convective loop. The thermal mass of the floor slabs helps attenuate the diurnal swings.

While thermal performance of the building is a major parameter of design, adequate distribution of daylight within the entire working zone of the building is a major criteria for design. This has been achieved through the domical structures designed above the light wells, which are evenly distributed throughout the building. Consequently, the design is thermally responsive to its climatic context and good daylight distribution is achieved, thereby minimizing the consumption of electricity.

Performance The building is under construction and hence no performance results are available.

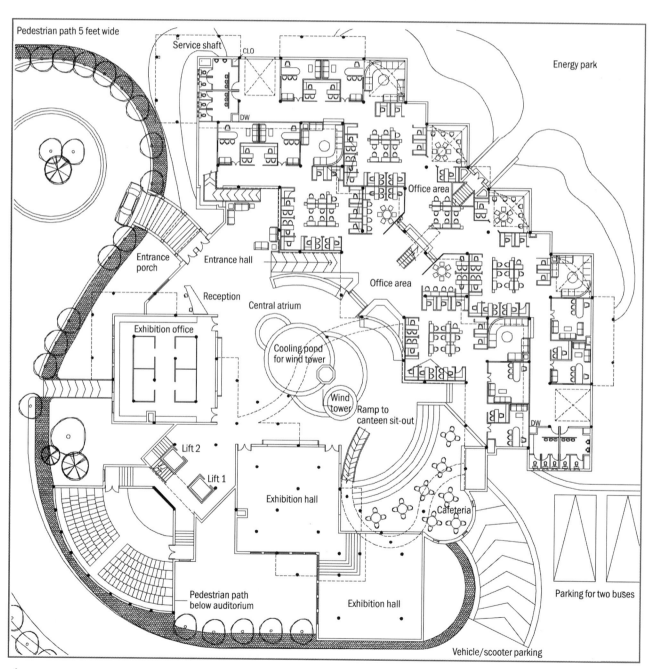

▲ Site plan of PEDA office complex, Chandigarh

PEDA office complex, Chandigarh

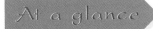

The PEDA office complex is located in Chandigarh on a practically square site with no major topographical variations. The structure is designed to achieve a climate-responsive building with a series of overlapping floors at different levels with interpenetrating large vertical cut-outs.

Project details

Building type Commercial (office building)
Climate Composite
Architects Arvind Krishan and Kunal Jain
Built-up area 7000 m²
Client/owner Punjab Energy Development Agency
Contractor Amarnath Agarwal and Sons, Chandigarh
Completion (expected) December 2001
Cost Not available

Design features

- Floors interconnected volumetrically to enable passive space conditioning of the entire volume of the building
- Large cut-outs for light and ventilation wells
- Building-integrated solar photovoltaics and solar water heating
- Winter heating by direct solar gain through roof glazing
- Summer cooling through wind tower
- Thermal mass of floor slabs moderates diurnal swings

Bidani House, Faridabad

Architects Arvind Krishan and Kunal Jain

A residential building that responds to climatic needs to provide comfort

Editor's remarks

In an urban context, with constrained site size and fixed orientation, the Bidani House demonstrates a climate-responsive design, which provides visual and thermal comfort round the year.

Very often it is stated that it is possible to design climatically responsive buildings on a larger site, but in most urban situations where the sites are constrained by their small size and fixed orientation, it is not possible to develop such a design. The Bidani House is a project that demonstrates a situation where a climate-responsive form and design was achieved in an existing urban situation with a fixed site size and orientation.

Faridabad, located in the 'composite climate' zone, has large climatic swings over the year, i.e. very hot and dry period of almost two and a half months and a colder period of a shorter duration. The hot dry period is followed by a hot humid, monsoon period of about two months with intervening periods of milder climate.

▲ Climate-responsive form to maximize heat loss in summer and heat gain during winter

The site

Located in Faridabad, near New Delhi, this house has been designed and built in the 'composite climatic context'. The site of about 1000 m² had a plan area in the ratio of 1:3 with the shorter side facing the road and oriented north.

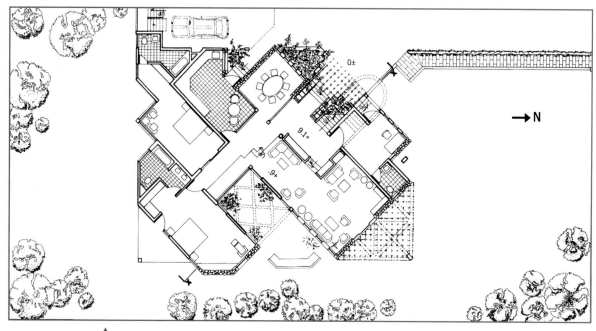

▲ Maximum exposure to south-east for living spaces and buffer spaces on the south-west to eliminate heat gains during summer

Planning in response to climate

The demand on building design was to respond to the extremes: eliminate (minimize) heat gain in hot dry period, maximize ventilation in hot humid period from zones/areas designed as heat sinks and maximize heat gain in the cold period. This has been achieved in this house entirely through the form and fabric of the building. A courtyard facing and opening onto north-east has been designed as a heat sink. The entire house form has been developed around the courtyard with all the main living spaces wrapping around it and having maximum south-east orientation that is the ideal exposure for this context. A large volume living space designed as a double height space is wrapped around the courtyard. Buffer spaces like the toilets and stores are located on the overheated south-western exposure to eliminate heat gain in summers.

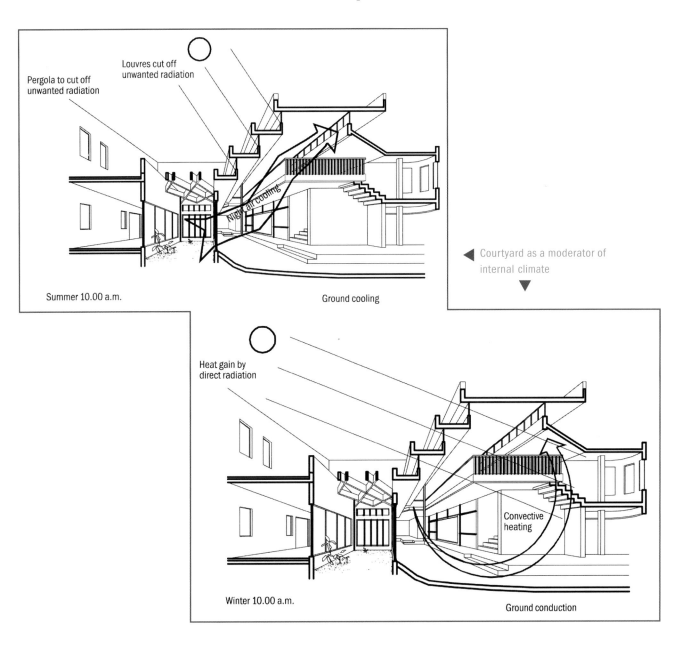

Courtyard as a moderator of internal climate

The three-dimensional form of the building is generated to eliminate or allow solar penetration according to seasonal changes. Large volume spaces and their coupling with the courtyard also allow good ventilation from the courtyard (the heat sink). The plan and three-dimensional form of the building has, therefore, been developed entirely in response to solar geometry.

Double-height living space for ventilation and daylighting

Diurnal swings in temperature are attenuated by judicious design and placement of thermal mass, utilizing local stone as the major material of construction.

The resultant building provides a comfortable environment with the temperatures, humidity, and airflow levels remaining in the comfort zone during all seasons of the year.

At a glance

Project details

Project name Bidani House, Faridabad
Site area 1000 m²
Climate Composite
Building type Residential
Architects Arvind Krishan and Kunal Jain

Design features

- House form developed around courtyard (acts as heat sink)
- Large volumes of spaces coupled with courtyard for ventilation
- Buffer spaces located on the overheated south-western exposure
- Form of the building allows solar penetration according to seasonal changes
- Pergola and louvres cut off unwanted radiation
- Local stone used as major construction material, which provides thermal mass for attenuation of diurnal swings in temperature

Transport Corporation of India Ltd, Gurgaon

Editor's remarks

An office building in the composite climate of Gurgaon with a climate-responsive built-environment to take advantage of seasons and thereby facilitating reduction in energy consumption.

Architect Ashok B Lall

An experience of all seasons as an aspect of a work environment

The corporate office building of the Transport Corporation of India has been designed to meet the demands of a modern office, with high level of environmental comfort, integration of systems to support information technology, with flexibility and adaptability for growth and change.

The building sits on a rectangular plot in an institutional area close to Delhi. Three stories of offices and a basement surround the central court. The basement houses building services and some work spaces.

▲ A view of the building highlighting the combination of mass and void, which has been carefully detailed out for catering to ventilation and daylighting

The building opens towards its entrance through a planted and shaded forecourt with a water pool. The orientation of all the interior spaces is towards the central court with the exception of the managing director's suite, which enjoys its own garden terrace on the top floor.

Architectural concept

The basic design strategy is inspired by the traditional inward-looking haveli plan. The central fountain courtyard acts as an environment generator for the office spaces opening towards it. The external skin is treated as a solid insulated wall with peep windows for possible cross-ventilation and higher windows for daylight. Selection of materials and systems of environmental control is prioritized in favour of sustainability and efficiency in energy consumption.

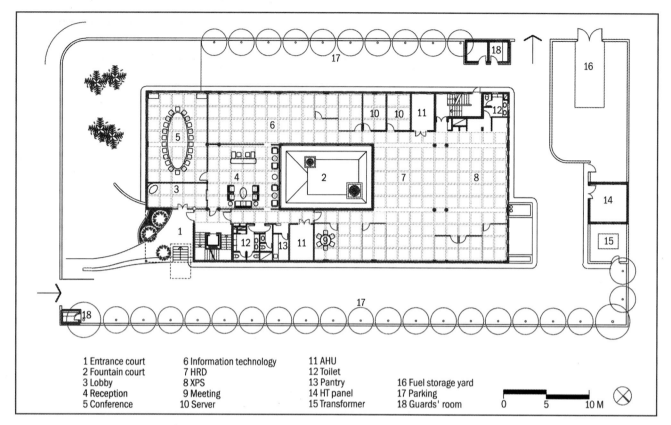

▲ Ground-floor plan of the Transport Corporation of India, Gurgaon

1 Entrance court	6 Information technology	11 AHU
2 Fountain court	7 HRD	12 Toilet
3 Lobby	8 XPS	13 Pantry	16 Fuel storage yard
4 Reception	9 Meeting	14 HT panel	17 Parking
5 Conference	10 Server	15 Transformer	18 Guards' room

Energy-saving features

Exposure

Being situated in a composite climate with climatic extremes, the building adopts a compact rectangular form and minimum height above ground to limit exposure to the external conditions. Openings on the external walls are designed for two separate functions: small peep windows at seating height provide for possible cross-ventilation and views out; larger windows at ceiling level are designed to distribute glare-free daylight across the office floor. Taking the daylighting function into account, the window area is minimized to 18% of the external wall area.

Both the entrance forecourt and central fountain court, towards which the building envelope opens out with greater transparency, have a structural framework which would provide support for shading screens to be stretched according to seasonal demands. The planting scheme along the edges of the site with tall evergreen (Silver Oak) trees provides another protective layer for the building.

Insolation

The orientation of the building is determined by the site. The small peep-windows, due to the deep reveal in which they are set, allow insolation in winter, cutting out the mid-summer sun by the shade of the reveal on to the glass. The large daylight windows house adjustable Venetian blinds in a double-window sandwich. The blinds are to be adjusted seasonally (thrice a year) by the building maintenance staff to control direct insolation and to reflect light towards the ceiling for distribution into the office spaces. The large glazed areas towards the central court and the entrance court rely on screens that will be stretched and gathered seasonally. The structural frameworks enclosing the courts provide the necessary support systems for the screens. It is planned that the screens would be works of art in themselves, which play upon opacity, translucency, reflection, and colour as ways of accentuating the experience of seasonal change.

Heat transfer

In principle, the building is a heavy mass construction insulated from the outside. Wall insulation is a 25-mm thick polyurethane foam protected by a dry red-stone slab cladding system. The roof insulation is 35-mm thick and has a reflective glazed tile paving cover to minimize sol-air temperature on the roof surface. The daylight windows provide insulation by way of tight-sealed two layers of glass with a Venetian blind installed between the two layers.

The glazing panels around the inner courtyard, however, are single glazed; it is anticipated that with the tall water fountain working, the courtyard temperatures would shift substantially towards wet-bulb temperature. This would considerably reduce heat load from the courtyard side during summers, and during spring and autumn would act as a heat sink. The choice of single glazing here evidently means savings in capital expenditure, considering the year-round operation of the fountain court.

Fountain court

The fountain court is an environmental device that seeks to combine the principles of physics, perception, and cultural psychology to produce an aesthetic language in which nature is reinstated as a beneficient force in architecture.

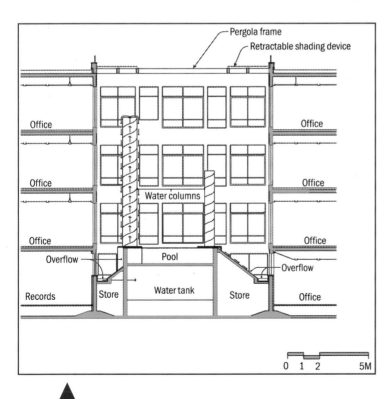

The fountain court is a generator of internal microclimate. It acts as a large heat sink during summers. The innovative section of fountain court helps in uniform daylight distribution even to basement areas

The fountain is designed as a visible object in which water can be seen and heard from all levels of the building, catching the light from above and in a variety of forms of movement. The gentle play of water and the pool conveys a sense of peace and tranquility. The use of white-textured concrete of the columns and white marble of the pool establishes its status, by association with tradition, as a work of art.

The fountain is a recirculating system in which a large body of water flows over extensive surfaces to maximize evaporation. The tall solid concrete columns of broad diameters over which the water trickles down the height of the courtyard, and the thin sheet that overflows the sides of the pool at ground level create a large heat sink and a body of air close to wet-bulb temperature. The white marble sides of the tank reflect the courtyard light into the basement work areas.

Interactive strategy for an air-conditioned building

Recognizing that climatic conditions range on both the cold and the hot sides of the comfort zone, building systems are designed to draw upon the external environment to supplement the air-conditioning system. Insulation is clothed on the outer surface of the building envelope giving a high thermal mass interacting with the interior spaces.

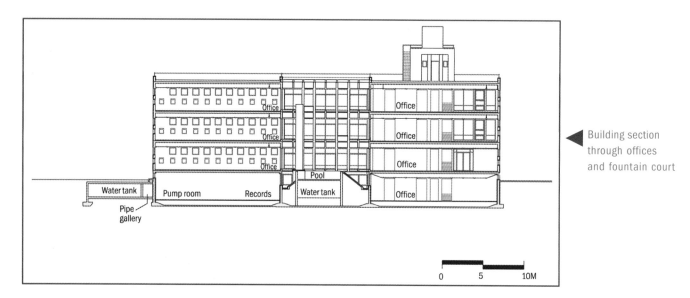

Building section through offices and fountain court

The air-handling system has provision for 100% filtered fresh air intake. Coolth can be stored in the building mass by night flushing during spring and autumn. Similarly, during early and late winter, when internal heat is to be rejected, fresh air would be drawn in, replacing the function of the chilling plant.

Absorption technology for air-conditioning

After a careful cost-benefit study, an absorption system chilling plant has been installed. Apart from not contributing to ozone depletion, the plant results in reduction of the capital expense of the electrical system, particularly its electricity generation back-up. This is of critical value in a state like Haryana, where due to acute electricity shortage the electricity generation back-up must cater to 100% of peak load. The absorption chillers run on a diesel-fired furnace. Electricity generation provides for illumination, office machines, and mechanical equipment.

Air distribution

Each of the office floors is served with two air-handling units. The allocation of areas handled by each unit is designed to balance out peak demands on each unit. This is done by responding to the orientation exposure of the building sides to the sun so that peak morning and afternoon loads are shared by the air-handling units even as the loads shift from the easterly faces of the building in the morning towards the westerly faces in the afternoon.

Control on air-conditioning loads

The primary-level control on external gains has been described under exposure and heat transfer; and internal gain is controlled by minimizing internal lighting loads. Air-conditioning standards are set according to the acceptance level of the office staff, rather than by any international norms, thereby resulting in significant energy saving. The system is designed based on the following parameters.

Outdoor	Summer 43.5 °C dry bulb temperature, 24 °C wet-bulb temperature (ignoring peak temperatures)
Indoor	Summer 24 °C dry bulb temperature (± 1 °C)

Also, circulation passages and ancillary function rooms have no air-conditioning. Toilets and pantries expel air to the outside at a minimal rate drawing relief air from neighbouring conditioned spaces.

Illumination

Daylight is the primary source of illumination. All work spaces receive adequate daylight, the maximum distance of a workstation from the daylight source being 5 m. The high windows on the external walls are designed to throw daylight deep into the office space. This is varied seasonally by adjusting Venetian blinds installed in the window sandwich to control glare and to modulate distribution. On the courtyard side, fabric screens would be stretched over the structural frame to respond to each season.

Artificial illumination is on the ceiling grid with compact fluorescent luminaries at 5 W/m^2 of floor area. Most of the office work is done on computers and working hours are generally limited to daylight hours. The illumination level offered by this system supplements daylight when necessary, and is comfortable for short working hours. It has been agreed that task light desk lamps will be provided on desks for elderly people and those who have late working hours. To provide visual interest and a feeling of brightness, occasional spot lights are provided to light up wall surfaces with paintings and other artwork.

Control of ceiling lights is in the hands of the building management staff. The control circuits for ceiling lights are arranged in zones running parallel to the daylight source so that they can be switched on progressively to compensate for variation in and/or falling daylight levels. It is proposed that these will be controlled by automatic timer switches with timing set for each season (with manual override for unusually cloudy weather).
A significant feature of energy saving is the economy of the building envelope.

Structural system and floor-to-floor height

A flat-slab system is adopted for floors and roofs. This minimizes the height required for accommodating air-conditioning and other services. With a clear ceiling height for office spaces at 2.65 m, the floor-to-floor height of the building is 3.5 m. This compactness of height means minimizing heat transfer through vertical surfaces of the external skin.

Restricting the building height to three stories was a deliberate choice. With maximum ground coverage, this pattern of planning consumes the total permissible floor area ratio with the least possible building height. The major advantage is that the energy consumed in transport of materials to heights during construction is minimized. Similarly the energy consumed in conveying water and diesel for the air-conditioning plant on the roof is minimized. A major gain is being able to eliminate the necessity of lifts. Only one 6-passenger elevator is provided for disabled or ill persons and for special guests.

Embodied energy

It is in the deployment of finishing materials of the building that some gains are affected by conscious choice. The criterion for choice of materials was that within the constraints of performance specifications demanded of the surface, the material should be chosen from the nearest possible source and should call for minimum processing towards converting or installing it. The external cladding is undressed split red Agra sandstone with precast concrete and terrazzo cills and jambs. For office areas, floors are pre-polished granite from Jhansi (the nearest source to Delhi). For service areas, it is Kota stone. The use of glass and aluminium is kept to the minimum possible.

Monitoring and automation

For the present, automation in the air-conditioning system is limited to the solenoid control valves and thermostats to regulate the flow of chilled water to the air-handing units and the switching on and off of the absorption chiller units; and for artificial illumination, the use of switches on timers. More sophisticated computerized automation systems were found to be beyond budgetary provisions and of doubtful cost-benefit. However, it is proposed to install a simple monitoring system for illumination and air-conditioning to help in rationalizing the systems management routines for the daily as well as the annual cycles of building use.

Performance survey and user feedback

The building is the corporate headquarters of the Transport Corporation of India. The air-conditioning requirement of the building is met by two 62.5 tonnes of refrigeration chillers. The space-conditioning requirements of the building are met by operating a single chiller for major part of the year. Two chillers are operated from mid-April through September only. The average electricity consumption is about 330 kWh per day. The office areas including the basement areas are very well daylit and artificial lights are not required to be switched on during the daytime. Water conservation is practised by using fountain and cooling tower discharge for gardening.

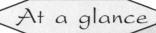

Project details

Site address No. 69, Sector 32, Institutional Sector, Gurgaon, Haryana
Architects A B Lall Architects
Climate Composite
Year of start/completion 1998/99
Client/owner Transport Corporation of India Ltd
Total built-up area 2750 m²
Cost:

Infrastructure (electrical, plumbing, HVAC, lift, fuel oil tank, pumps and tubewells)	Rs 24 million
Civil, false ceiling, strong rooms, steel pergola at entrance	Rs 30.7 million
Landscaping	Rs 0.35 million

Design features

- Inward-looking compact form, with controlled exposure
- Two types of windows designed: peep windows for possible cross-ventilation and view, the other being for daylighting
- The courts towards which the building has more transparency have structural framework to provide support for shading screens
- Landscaping acts as a climate modifier
- The window reveals of the peep window cut out summer sun and let in winter sun
- Adjustable Venetian blinds in double window sandwich to cut off insulation and allow daylight
- Polyurethane board insulation on wall and roof
- Fountain court with water columns as environment moderator
- Building systems designed so as to draw upon external environment to supplement the air-conditioning system
- Eco-friendly absorption technology adopted for air-conditioning
- Careful planning of air distribution system
- Air-conditioning standards set by acceptance level of office staff and not by international norms
- Energy-efficient lighting system and daylight integration with controls
- Optimization of structure and reduction of embodied energy by use of less energy-intensive materials

Temperature check for Gurgaon

(see Appendix IV)

Cold	Cool	Comfortable	Warm	Hot	
	•				January
	•				February
		•			March
			•		April
				•	May
				•	June
			•		July
			•		August
			•		September
		•			October
		•			November
	•				December

Add the checks

	3	3	4	2

Multiply by 8.33% for % of year

Heating	25
Comfortable	25
Cooling	50

SOS Tibetan Children's Village, Rajpur, Dehradun

Editor's remarks

Low-cost, low-maintenance construction designed to ward off harsh winds and facilitate solar access and ventilation. Simple construction techniques have been adopted to provide insulation at a minimal cost. Landscape planning has been carefully done to provide shelter from cold winter winds and access to winter sun.

Architect Ashok B Lall

Innovative use of passive solar techniques, with carefully detailed out building envelopes and components for a low-cost children's village

The site for this residential campus measures 14 300 m² and is located at Rajpur in the Mussoorie foothills. There are 15 family cottages that provide home and education to 225 children, residences for the director and co-workers, community facilities and a place of worship. The hilly site, a part of which was a mango orchard, slopes generally towards the south.

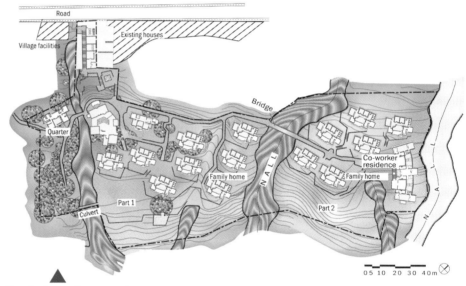

The site has a slope towards south, which allows solar orientation of the cottages

Concept The emphasis is on simple low-cost, low-maintenance construction. The primary strategy is to provide protection from harsh winds from the north-east and to provide solar access. The outdoor, used extensively by all the residents, is also designed as 'habitable' space.

◀ Habitable outdoor space

Outdoor spaces The planning and plantation schemes are combined. The large playground is in a wind-sheltered zone with clear winter sun access, while the existing fruit orchard doubles as a shaded playground. Smaller pockets between buildings, benches under the shelter of trees and low walls to sit or slide on are also planned as places for informal recreation.

Building plans All buildings are adjusted to the terrain and face south. Where buildings are two-room deep, the section is stepped and the plan joggled to allow the access of some winter sun from the northern side. Similarly, when buildings are located one behind the other on a slope, their levels are adjusted to receive winter sun. All the homes have an outdoor terrace on the south, designed as a natural extension to the front verandahs and function as adjacent outdoor living spaces.

The toilet block that occupies the north-east corner of the family home building acts as a buffer against cold winter winds.

The sections of the buildings provide for clerestory windows that allow some winter sun from the north and also provide for cross-ventilation in the summer.

▲ Buildings are adjusted to terrain and face towards south. The clerestory windows bring winter sun to the rooms at back

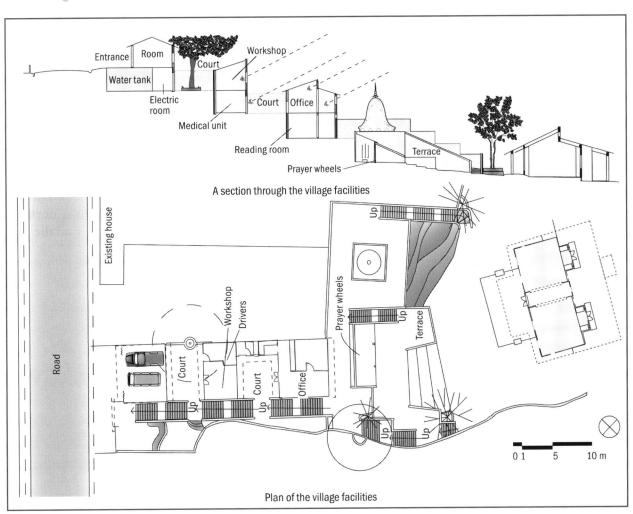

▲ Plan and section through a building showing solar orientation

Building fabric Three simple measures are incorporated to improve the thermal performance of the buildings.

- The 100-mm thick RCC roof slab is clad with hollow terracotta tiles to provide some insulation on the outer side of the thermal mass. This improves the thermal dampening characteristics of the roof.
- The walls are finished with light-coloured stone aggregate plaster. This reduces sol-air temperatures on the wall faces.
- The bedrooms have solid timber board shutters on the inner side (instead of curtains). These are shut during winter nights and summer afternoons to improve insulation and ward against infiltration. All window shutters are designed with a double rebate to reduce infiltration.

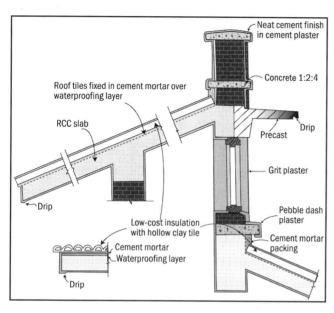

▲ External roof insulation with hollow terracotta tiles to improve thermal dampening characteristics of the roof

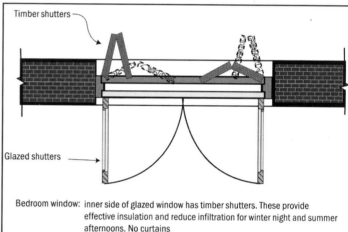

Window detail for bedrooms ▶

Bedroom window: inner side of glazed window has timber shutters. These provide effective insulation and reduce infiltration for winter night and summer afternoons. No curtains

At a glance

Project details

Site address SOS Tibetan Children's Village, Rajpur, Dehradun, Uttar Pradesh
Building type Institutional
Architect Ashok B Lall
Year of start/completion 1997–1999
Climate Composite
Client/Owner Tibetan Homes Foundation, Mussoorie
Total project cost Rs 45.8 million

Design features

- Aims to protect houses from harsh winds and to provide solar access
- Building plan and section maximizes solar access during winter and ventilation in summer
- The toilet block in north-east corner of family homes acts as buffer against cold winter winds
- Roof insulation by hollow terracotta tiles
- Sol-air temperature on the wall faces reduced by light-coloured stone aggregate plaster
- Bedrooms with solid timber board shutters on inner side for insulation
- Windows with double rebate to reduce infiltration

Temperature check for Dehradun

(see Appendix IV)

Cold	Cool	Comfortable	Warm	Hot	
	●				January
	●				February
		●			March
		●			April
			●		May
			●		June
			●		July
			●		August
		●			September
		●			October
		●			November
	●				December

Add the checks

	3	5	4	

Multiply by 8.33% for % of year

Heating	25
Comfortable	42
Cooling	33

Redevelopment of property at Civil Lines, Delhi

Editor's remarks

This cluster of four residential units in the high density area of Delhi demonstrates that site constraints do not deter incorporation of passive solar measures and these constraints can be effectively tackled by innovative planning and detailing. Interactive courtyard is a novel concept, which can be used in a composite climate.

Architect Ashok B Lall

Passive devices and innovative construction methods are a winning combination in these residential units

The large open plots of land with bungalows that constituted Delhi's Civil Lines are now being sub-divided and redeveloped to provide more upper-middle income housing. The resultant urban structure has the nature of a close-grained texture with low-rise, high-density housing. In some ways it can be seen as a mutation of the North Indian city form. This project explores the possibility of responding more deliberately to climatic factors in a dense setting.

Traditional streetscape: low rise high density housing which can be seen as a mutation of the North Indian city form

The project consists of four residential units built-to-edge on a street. The houses on the north face of the street are courtyard houses leading towards gardens on the south side. The houses on the south side of the street have their gardens on the north side and are linear. These are all large single family houses, two to three stories high. This enables the sections of the buildings to be designed integrally for enjoying the winter sun. The passive devices that interact with the external elements are given a central place in the architectural language of the buildings.

Passive solar features

Insolation

The general orientation of the buildings is east–west, making most of the window openings fall in the north and south faces. The courtyard houses, because of their rather square proportions in plan, have faces towards the east and west as well. The windows on these faces look into narrow protected alleys or the small courtyard between the two houses. Retaining the wall of the original double-storey building which had lined the street shades the alley space on the west.

Ground-floor plan ▶

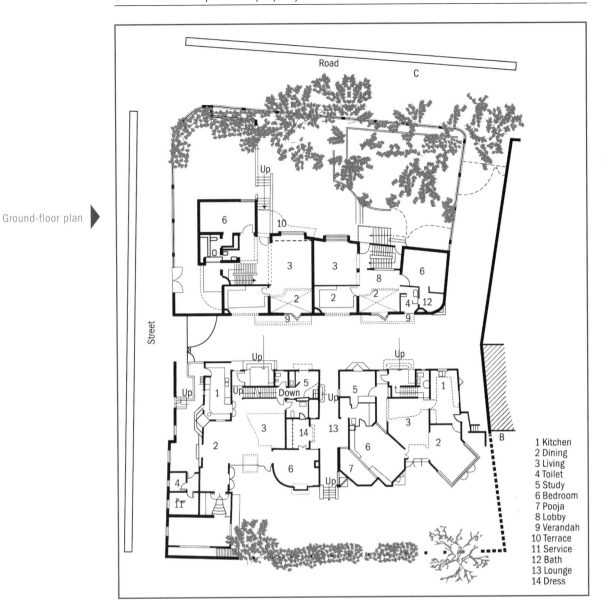

1 Kitchen
2 Dining
3 Living
4 Toilet
5 Study
6 Bedroom
7 Pooja
8 Lobby
9 Verandah
10 Terrace
11 Service
12 Bath
13 Lounge
14 Dress

For the linear houses on the north side, the width of the driveway that separates the two rows of houses is just enough to enable winter sunshine enter the first-floor windows. The sections of these houses are designed with a cut-out such that the winter sun is brought into the living/dining space – the heart of the house – on the ground floor. Terraces on the second floor have skylights that again admit winter sun into the first-floor rooms on the north side of the house.

Insulation

Roofs are finished with broken China mosaic. The roof construction sandwich contains 30-mm thick polyurethane board insulation above the RCC slab.

For the courtyard houses, the western wall of the upper floor, the east and west walls of the courtyard roof, and the water tank walls are insulated using an innovative construction sandwich. Insulation board is pasted on to the outside of a 115-mm thick brick wall and held in position with panels of terracotta *jalis* whose cavities are rendered with cement sand mortar. Their resultant construction expresses the special nature of the wall as a decorative textured surface.

The courtyard roof

The main climate-responsive device is the roofed courtyard of the two courtyard houses. The hipped steel frame roof is clad with a sandwich panel of fixed frosted glass with a tinted film for the most part with a small south portion through which the sky is visible. This is underslung by a pair of *razais* (quilts) which can be pulled across to cover the underside of the roof (for insulation) or allowed to hang down vertically (to allow heat transfer). Above the roof is another frame in *chics* (bamboo severs), which can similarly be opened to shade the roof or rolled up to catch the sun.

The ridge of the roof is a channel from which water overflows on to the thin roofing membrane of stone and glass. Some water would evaporate and excess water is collected at the foot of the slope and recirculated. This makes the roof a large evaporative cooler over the central space of the house. All rooms communicate directly with this central space. This method of evaporative cooling will supplement a conventional evaporative cooler. And in the hot-humid period of July to August, this would give considerable cooling, when the conventional evaporative cooler is no longer effective.

Continuous shading devices on south wall and window protect south wall and window from direct insolation during summer

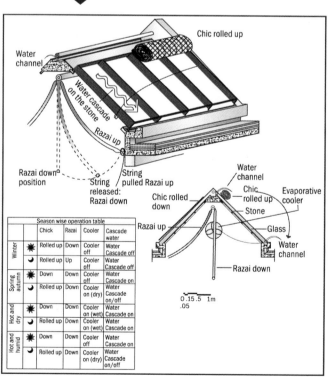

Interactive courtyard roof for climate control of the entire house in all seasons

The operation of the roof component (*chic*, water, and *razai*) is to be adjusted from winter to summer and for day and night.

The roof provides for (1) shading from outside / insulation from inside, (2) roof evaporative cooling, and (3) direct (through clear glass) and indirect (through frosted glass) radiation. The glass also acts as a night radiator in summer albeit a rather inefficient one.

The dominant portion of the roofed courtyards with their quilts of mirrors and colourful cloth, the *chics* and the possibility of visible monsoon and night sky would become a strong aesthetic experience responding to the rhythm of seasonal cycles.

Infiltration

All windows communicating with the outside are designed with double rebates. So are all external doors with the traditional *chowkhat* (four-sided frame). This controls infiltration of both cold/hot winds as well as dust, which is a major household maintenance concern.

Air-conditioning	The office basement is supplemented by a 5 TR (tonnes of refrigeration) 'split' air-conditioning unit (to condition an area of 120 m²), which is used for approximately 60% of the working hours from May to August.
Conventional evaporative cooler	A conventional mosquito-proof evaporative cooler is housed on the roof blowing through the sidewall of the courtyard roof. This is an ideal system for the hot-dry season and for night flushing during the humid season as well as during autumn and spring.
Wind-driven evaporative cooler	The West House takes advantage of the prevailing north-westerly hot winds that blow during the hot-dry seasons. A vertical screen tower is built on the west wall with evaporative *khus* pads on its outer surface fed by a water pump. The inner side has adjustable windows opening into the adjacent rooms. The natural wind pressure will drive air through the wet *khus* pads into the adjacent rooms.

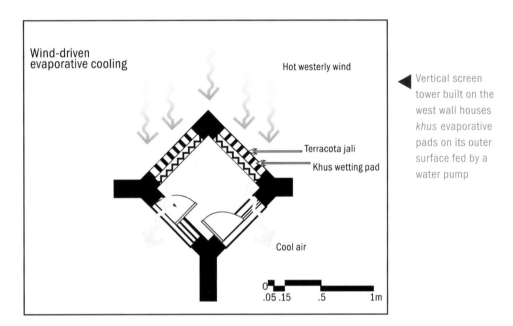

Vertical screen tower built on the west wall houses *khus* evaporative pads on its outer surface fed by a water pump

Basement	The basements are designed as multipurpose spaces with adequate natural light. The basement of the west courtyard house is being used as an architect's office. The artificial illumination is largely by way of compact fluorescent desk lamps.
Performance	These houses are non-air-conditioned buildings. However, there is a provision for installing room air-conditioners in the bedroom. And there is a 5-TR air-conditioner for the basement office (120 m²). The design strategy is not directed to predictably control the thermal performance of the indoor spaces—rather it is to integrate some features in the buildings that would shift its performance substantially towards comfort. The buildings are just completed and they will be tested for performance over the year. The courtyard roof is a crucial element of the buildings and a lot depends on its effectiveness.

With correct operation of the quilt during winters, night-time comfort, say 18 °C minimum, can be assured in all rooms.

For the basement offices, the air-conditioners were switched on 15 May 2000, and over the summer months the air-conditioner was used for about 60% of office time. For the ground and first floors, comfort can be obtained during the dry-hot season and during spring (March–April) and autumn (September–October) using a combination of direct and roof evaporative cooling with ceiling fans. During the monsoon season (July–August) roof evaporative and cross-ventilation will give moderate comfort during daytime and comfort at night. But in the pre-monsoon hot-humid season (June–July) with little breeze, murky sky, high humidity and temperature, one will have to resort to air-conditioning for comfort.

At a glance

Project details

The project explores the possibility of responding more deliberately to climatic factors in a dense setting.

Building type Residential
Climate Composite
Location It is located in the Civil Lines area of Delhi where large open plots of land are being subdivided and redeveloped to provide more upper-middle income housing
Architect Ashok B Lall
Built-up area 1687 m²
Year of completion 1999

Design features

- Orientation of the building to cut off solar insolation during summer and let in winter sun
- Design of sections to let in winter sun into the first-floor rooms on the north side of the house
- Terraces with skylights that admit winter sun
- Insulated walls using innovative construction sandwich
- Sunshading reduces heat gain
- Courtyard design
- Roof finished with China mosaic and is insulated using 30-mm thick polyurethane board insulation above the RCC slab
- The courtyard roof is the main climate-responsive device acting as a large evaporative cooler over the central space of the house. All rooms communicate with this space
- Conventional mosquito-proof evaporative cooler housed on the roof
- External windows are designed with double rebates
- The west house has a wind-driven evaporative cooler

Temperature check for Delhi

(see Appendix IV)

Cold	Cool	Comfortable	Warm	Hot	
	●				January
	●				February
		●			March
			●		April
				●	May
				●	June
			●		July
			●		August
			●		September
		●			October
		●			November
	●				December

Add the checks

	3	3	4	2

Multiply by 8.33% for % of year

Heating	25
Comfortable	25
Cooling	50

Integrated Rural Energy Programme Training Centre, Delhi*

Editor's remarks

This building provides comfortable working conditions round the year, primarily due to a bioclimatic approach to design, judicious use of advanced passive condition techniques, and renewable energy systems.

Architect Manmohan Dayal

A building in composite climate where the room temperatures in summer are lower by 6–7 °C, and the winter temperature is raised by about 5 °C than in a conventional building.

Situated off the GT Karnal–Delhi road in Bakoli village, Alipur block, Delhi, the training centre constructed for the IREP (Integrated Rural Energy Programme) is located on a 2.5-hectare site. Originally an agricultural land, the site was being used as a brick kiln.

▲ The guest house at IREP with the energy park in front

The plan of the main office block of the IREP showing use of courtyards as micro climate modifier. Also shown is the network of earth air tunnel system. A 2.5-m wide and 1.2 m high earth berm runs round the building, which helps in moderating internal temperatures

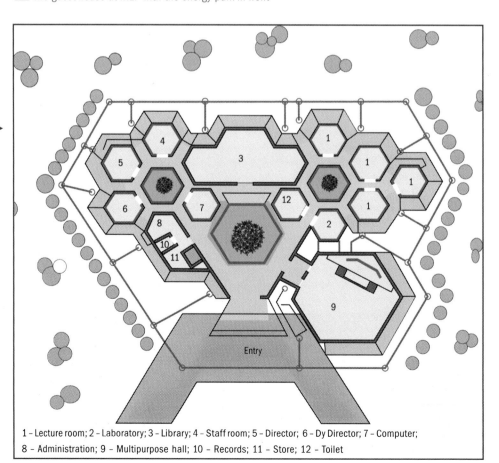

1 - Lecture room; 2 - Laboratory; 3 - Library; 4 - Staff room; 5 - Director; 6 - Dy Director; 7 - Computer; 8 - Administration; 9 - Multipurpose hall; 10 - Records; 11 - Store; 12 - Toilet

*The source of this chapter is *Architecture + Design*, **Volume IX** (Number 3), May–June, 1992 pp. 30–32

The general objectives of the project were
- integration of advanced solar passive architectural systems
- simplicity of operation and maintenance
- integration of climatic consideration in the form of the building, construction, and landscape.

The initial requirements were for four lecture theatres of 40 m² each, a library, an auditorium (having a capacity of 300 seats), a laboratory and computer room, administrative offices, and hostel facilities for 50 persons. Further, residential accommodation was provided for employees. Later, a guesthouse was also included.

Passive solar concepts

The several measures adopted to reduce the thermal heat loads in the complex are outlined below.

Controlling penetration of direct solar radiation

The south side is protected by *chajjas* (overhangs) of appropriate depth, which prevent the heat of the summer sun from entering, yet allow the winter radiation to penetrate within. A combination of horizontal and vertical louvres prevents the heat from the south-east/north-west sun, and at the same time provides adequate cross-ventilation.

Reducing heat conduction through roof and walls

The amount of heat conducted into the structure through the building fabric is dependant on the thermal resistance of the material and its heat storage capacity. Discomfort is caused by solar radiation, which is absorbed by the outside surface and transmitted through the roof or walls to the inside surface. The following steps have been taken to minimize conduction.

Shading
The complex is built around 3 courtyards with 75% wall area shaded by 2.1-m-wide corridors. A further 10% of this is north-facing (which is mostly shaded). Deciduous trees on the periphery of the building along the east and west orientation provide further shade. Creepers along the walls add to the insulation and help in evaporative cooling.

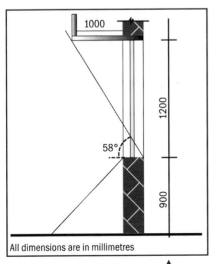
Shading of walls – South section

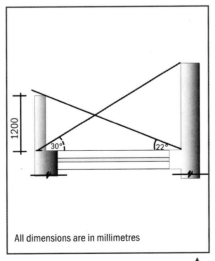
Louvre design of north wall typical plan

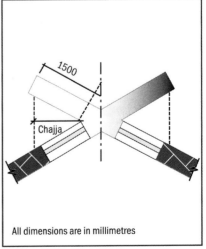
Louvre design of south-east wall typical plan

Insulation

It is essential to insulate the roof as it receives as much as 50% solar radiation in summer. A traditional cost-effective system has been used where the entire roof is covered with densely packed inverted earthen pots laid in mud phuska. This arrangement provides an insulation cover of still air over the roof, which impedes heat flow within the building.

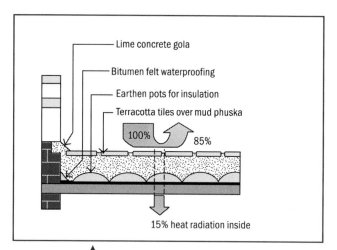

Roof insulation

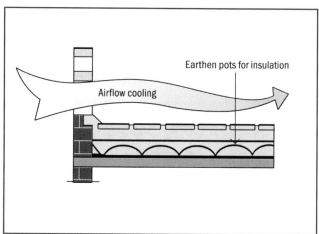

Parapet detail

Thermally massive wall and cavity wall

Designed for daytime use, the Training Centre has 345-mm thick exposed brick walls that have a time lag of about 8 hours, and an overall heat transfer coefficient (U-value) of 1.9 W/m² °C. Cavity walls (230-mm brick wall + 35-mm cavity + 115-mm brick wall) envelop the computer centre to give an additional air gap insulation of 35-mm (U-value 1.2 W/m² °C and a time lag of 12 hours).

Earth berming

Direct contact with the earth creates better comfort levels than evaporative cooling, as it does not increase the humidity within the rooms. While it was not possible to build a totally earth-sheltered building, an earth berm, 2.5-m wide and 1.2-m high, runs along the external periphery of the structure. The induction of earth mass to the thermal mass of the building reduces fluctuation in the thermal load besides acting as a form of shading and insulation.

Control of ambient temperature and micro climate

The excavated areas of the site are converted to a water body, which act as a collection centre for rainwater and surface drainage. Hard surfaces have been minimized to roads and the rest of the area is landscaped. Within the Training Centre, the landscaped courtyard with a tree each in the smaller courtyards and a fountain in the larger, help in reducing the ambient temperature. Narrow passages between each room act as a low-pressure area forcing more air into the neck thereby improving the airflow within.

Exploitation of wind and earth for cooling

The temperature of the earth 4 m below ground level remains constant throughout the year and is equal to the mean annual sol air temperature of earth. The earth can be used as a heat source or a sink for heating/cooling air in underground pipes as the earth–air heat exchanger system utilizes the stable temperature and large thermal capacity of the earth. Air that is

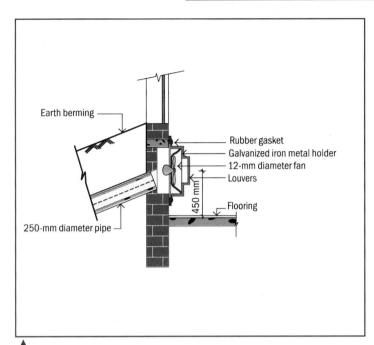

Details of ducting of earth air tunnel

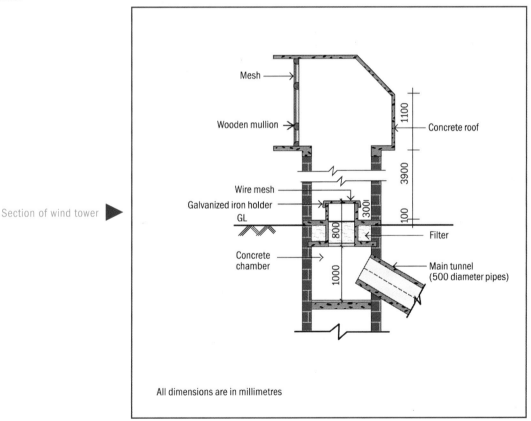

Section of wind tower

passed through the pipes 4 m below ground level is cooled in the summer and heated in the winter. The amount of heat exchanged between the air and the surrounding soil depends on various parameters of surface area, the length of pipes, velocity of air, etc. In the Training Centre, a 500-mm diameter pipe of length 50 m connects the wind tower fixed with a blower to the 300-mm diameter ring pipe. The connecting pipes (of 250-mm diameter) enter the rooms at below-sill levels from under the berm. Initially, it was proposed to use small blowers/fans of 300-mm diameter to blow air into the room. Unfortunately, the blowers could not suck sufficient air because of lack of space behind them. Finally, a 7.5-kW blower was fitted in the wind tower—a poor substitute, because the diameter of the pipes proved to be inadequate for the volume of air being pushed in at high velocity. The earth–air tunnel is sparingly used during May and June and is not effective due to its design limitations.

Installed renewable energy systems in the buildings and in the energy park

A 300-W roof-mount solar photovoltaic plant with battery back-up powers the computers in the building. Solar water heating requirements in the hostels are met by solar water heating system. The energy park within the complex has demonstration units for various renewable energy systems.

A greenhouse, solar still, solar refrigerator, smokeless *chulha*, gasifier unit, and a biogas unit have been put up for demonstration. The biogas generated from the biogas plant is used for cooking in the hostels. The solar refrigerator cools drinking water during summers. A solar pump has also been put up for water pumping purposes.

Performance Although no data have been collected by the Delhi Energy Development Agency on the performance of the various passive systems used in improving the thermal comfort level, it can be safely said that the summer daytime room temperatures within the Centre are lower by 6–7 °C, and the winter temperature is raised by about 5 °C compared to a conventional building. The users are comfortable round the year and do not feel the need for additional space-conditioning during summer or winter.

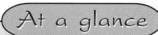

Project details

Project Integrated Renewable Energy Programme Training Centre
Location Delhi
Climate Composite
Architect Manmohan Dayal
Energy consultants N K Bansal and Muthu Kumar
Contractor M/s Bhagat Ram And Sons
Built-up area 7766 m²
Approximate cost Rs 20 million
Date of completion March 1990

Design features

- Control the penetration of direct solar radiation by proper shading
- Reduce heat conduction through roof and walls by insulation and earth berming
- Control ambient temperature and micro climate by landscaping
- Exploit wind and earth for cooling by earth–air tunnel system
- Meet partial power and hot water requirements through renewable energy systems in the building
- Energy park with demonstration units for several renewable energy systems

Temperature check for Delhi

(see Appendix IV)

	Cold	Cool	Comfortable	Warm	Hot	
		•				January
		•				February
			•			March
				•		April
					•	May
					•	June
				•		July
				•		August
				•		September
			•			October
			•			November
	•					December

Add the checks

	3	3	4	2

Multiply by 8.33% for % of year

Heating	25
Comfortable	25
Cooling	50

Tapasya Block (Phase 1), Sri Aurobindo Ashram, New Delhi

Editor's remarks

This hostel building for the ashramites, requiring medium-level equilibrium in terms of thermal and visual comfort, uses simple but effective passive solar architectural interventions. Use of building-integrated solar water heating system is an innovative feature.

Architect Sanjay Prakash

Application of passive solar techniques and renewable energy systems in a building with medium level of equilibrium

The Aurobindo Ashram Trust (Delhi branch) is a charitable society doing pioneering work in various fields, notably education, philosophy, and culture. The use of non-conventional energy techniques for various purposes is both economically and spiritually important for the philosophy of the Ashram. There was a large degree of participation by the ashramites for the construction.

This is the first phase of an unusual building complex which shall grow to form offices, swimming pool, library, dining, meditation, and residential facilities. In the first phase, it is planned to house around 80 people – students, visitors, and ashramites – in a hostel-like block.

The rooms are laid out around 3 hexagonal courtyards. The building as it grows will enclose one large courtyard around which there will be 12 such small duodecagonal blocks. The planning grid is a space-filling grid composed of squares and equilateral triangles. The courtyards are surrounded by overlooking corridors, around which the square grids are divided into two rooms and the triangles divided into two sets of toilets and dressing areas. Except on ground floor, each room gets a balcony on the outer faces.

Materials, techniques, and methods

Rough white finish

The building is finished in a permanent white slate, which is rough and, therefore, has good emissivity. Being light in colour, the absorptivity is poor.

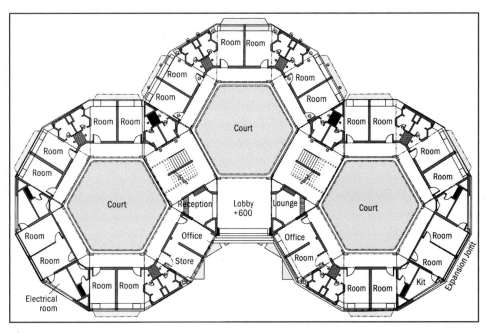

▲ The plan of the hostel is duodecagonal in shape centred around hexagonal courtyards. The courtyards along with carefully designed ventilators and windows aid in cross-ventilation

Light-coloured permanent finished walls have high emissivity and poor absorptivity. Deep recessed windows reduce solar gains

Deep recesses

Though the building is not oriented because of its multifaceted geometry, it has been provided with deeply recessed large openings on each room towards the outside.

Selective cross-ventilation

Each room is provided with an operable ventilator above the door on the 'inside' and large operable glazed and opaque shutters on the outside. Both have fixed insect-netting with inward opening glazing. The sills are low in height to be able to get good air flow even at the level of the bed. This feature has worked well, and fans are not required until April. The rooms can be opened up in the summer night and closed up with shutters and heavy curtains during the day to create temperatures, which average slightly lower than ambient condition.

Vegetation

Lowering of temperatures is also made possible by the densely vegetated surroundings of the campus in which this building is located. The selectively cross-ventilating air is drawn from the grassy and vegetated areas.

Courtyards

The inside courtyards are small and interlinked to encourage both cross-ventilation in rooms as well as ventilation between them.

Daylight

All rooms have an upper fixed glazing to encourage 'daylight-only' usage during the day. Small table-level windows are provided on study desks. Built-in window reading benches are placed below the main windows on the outer face of the ground floor rooms (which do not have balconies or terraces).

Installed system *Architecturally integrated hot water system*

Four thermosiphonic flat-plate solar collector systems for hot water are provided. There is no electrical back-up. The total capacity is 2400 litres/day (about 30 litres/person-day). Two systems, feeding the north-side rooms, are optimally tilted at 45 degrees. The other two systems, feeding the south-side room, are vertically mounted on the parapet wall facing south or ±30 degrees east or west of south. The building section especially at the terrace level was partially determined by solar hot water constraints. All systems are masonry integrated and mounted on walls or RCC (reinforced cement concrete) slabs. All tanks are neatly placed over toilet pipe shafts within room-like structures. This reduces balance-of-system costs while providing better aesthetics in comparison to retrofit systems. Further, the vertical systems display lesser dust collection (important in a polluted city like Delhi), lesser glass breakage and complete freeing up of terrace space, thereby offsetting the 25% lower performance for unit area over a winter season. The terrace is used for washing and drying of clothes. This system can be adapted to multi-dwelling multistorey buildings without the need to use terrace space.

▸ Parapet wall with integrated solar hot water panels. Also seen is the structure, which houses the hot water tank

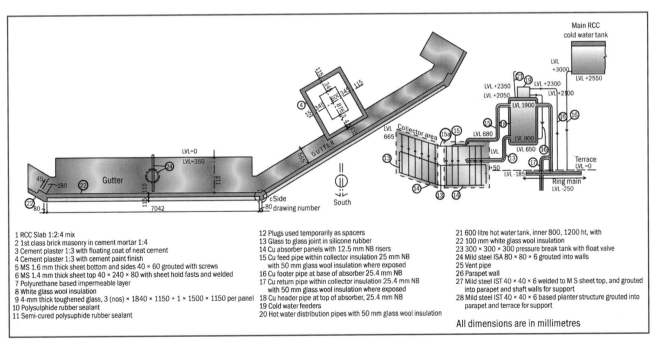

▲ Details of architecturally integrated solar water heating system (type B). Various details highlight integration of solar hot water collectors with parapet wall, hot water tank details, and piping details *(See the next figure for further details.)*

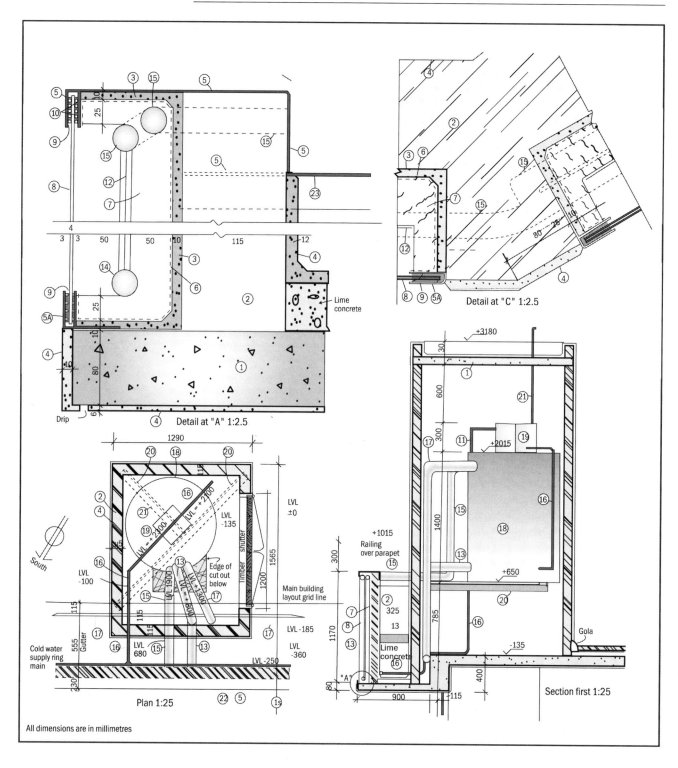

Electricity conservation Electricity is sparingly provided all over, with low voltage corridor lights, two lights, and one fan per room with three lights in the attached toilet and dressing areas. No 15 Ampere sockets are required. Even with incandescent lights, the overall installed power load is to the tune of $10\,\text{W/m}^2$, nearly all of it as lights and fans. Electrical energy conservation is partly due to the time table of the residents matching daylight patterns. The clothing and food habits also reduce the amount of cooling that is required. No power is used for cleaning, hot water, washing of clothes, toilets or any other function except (cold) water pumping.

Load-bearing structure The three-storey building was designed in load-bearing brickwork and was without beams and columns, reducing the consumption of concrete and steel.

Performance The solar hot water system has worked extremely well. There is no back-up electrical use for heating water. However, the large glass covers do break occasionally due to thermal expansion. Therefore, integrated collectors should also be made in small packs and further have a ledge in the front for ease of repair in case of problems. Both these changes are being incorporated in the next building of the Ashram.

The thermal behaviour of rooms is satisfactory for the lifestyle of the Ashram dwellers and the building uses no devices other than ceiling fans.

Economics The use of self-managed construction and use of load-bearing construction led to cost reduction by about 20%. The hot water system was additional though it would pay for itself in about two winter seasons even without grants.

Vertically-mounted collectors can thus be expected to give (for approximately 12% increase in capital costs) architectural integrated solar hot water systems along with great aesthetic gain, free terrace, less dust addition, less glass breakage, less maintenance cost.

Table I Effect of architectural integration on the economics of solar hot water system for New Delhi (winter solar water heating – passive model)

Configuration	Normal system (optimum tilt)	Integrated system (optimum tilt)	Integrated system Vertical
Collector area required	A_c	A_c	$4/3\,A_c$
1 Collector costs (x)	100.0	100	133.3
2 Balance of systems costs of passive systems at half of normal collector costs	50.0	50	50.0
3 Cost savings due to collector integration as a percentage of collector cost (reduction in back cover and supports)	–	6 (@6%)	14.6 (@ 11%)
Total costs (1 + 2 – 3)	150.0	144	168.7
Increase (decrease)%	0%	(4%)	12.5%

At a glance

Project details

Site location New Delhi
Project description Institutional hostel
Climate Composite
Design team Sanjay Prakash, Anop Singh Rana, and Manoj Joshi
Consultants C L Gupta and Atam Kumar (solar hot water)
Project period 1990–1993
Size 3000 m² covered area in a large campus
Building/construction type Load-bearing fired brick structure with RCC slabs, finished in white slate
Client/owner Sri Aurobindo Ashram, Delhi Branch Trust
Sponsor Ministry of Non-conventional Energy Sources (hot water system) and Sri Aurobindo Ashram, Delhi Branch
Builder/contractor Owner-managed construction

Design features
- Rough white finish on the building exterior for high emissivity and poor absorptivity.
- Deep recessed window offset the negative effects of rooms that are not oriented as per solar geometry
- Properly designed windows and ventilators and interconnected courtyards aid in cross-ventilation
- Dense vegetation modifies ambient conditions
- Window design encourages only use of daylighting in the rooms during daytime
- Building integrated solar water heating systems
- Load-bearing structure reduces embodied energy by reduced consumption of concrete and steel

Temperature check for Delhi

(see Appendix IV)

Cold	Cool	Comfortable	Warm	Hot	
	●				January
	●				February
		●			March
			●		April
				●	May
				●	June
			●		July
			●		August
			●		September
		●			October
		●			November
	●				December

Add the checks

	3	3	4	2

Multiply by 8.33% for % of year

Heating	25
Comfortable	25
Cooling	50

Residence for Sudha and Atam Kumar, Delhi

Editor's remarks

Designed for the composite climate of Delhi, this residential house demonstrates use of passive architectural techniques for achieving thermal and visual comfort in all seasons. Architectural integration of renewable energy devices is an important design feature. The installed electrical systems have also been carefully chosen so that the building requires minimal energy for its operation.

Architect Sanjay Prakash

A residential house in a composite climate responding to varying seasonal needs and maintaining comfort levels without any energy-intensive systems

This is the inaugural project in a series of EASE (environmentally appealing and saving energy) houses. EASE is a concept promoted by Cdr. C P Sharma (Retd) as a commercially viable method to propagate energy-efficient houses. This particular project was sold to a solar energy device manufacturer and consultant, who, interestingly, was also the energy consultant for of the building.

The house is a duplex structure with a living room, kitchen, three bedrooms, and a study. There is an outhouse with a servants' unit and office.

Materials, techniques, and methods

Site selection

From within the overall development, the site selected had a road to its south and west. The longer side of the plot faced south. The road on the south was the wider one, reducing possibilities of shadows from the south in the future.

 South elevation of the building with large shaded openings and solar chimneys as predominant architectural feature. Solar water heating collectors are integrated with the south facade

Planning of built volumes

The house was planned as a duplex to further reduce its footprint on the plot and was shifted as far north within the site, to the extent feasible, to allow for a large south open area.

South orientation

The house was oriented south in the sense that every habitable room has a liberal south exposure. Only entries, toilets, and staircases are without direct south orientation. The three bedrooms and the living room all have large south glazing for winter heat gain with proper overhang protection for prevention of summer heat gain. The south overhang soffit level is higher than the window lintel level so as to ensure that even a part of the windows is not shaded in winter. All winter heating needs can be met by the south glazing.

Insulation and mass

Expanded polystyrene (25-mm thick) insulation is provided near the outer surface of the walls so as to retain the mass of the wall acting in tandem with the internal space. Likewise, asbestos powder (40–80-mm thick) insulation is provided over the roof slab. Both these provide for a highly inert house with high thermal mass and high insulation.

Reflective finish

The walls are clad in white sandstone providing a textured and reflective finish. The roofs are finished in white terrazzo making for good terraces to sit out on as well as excellent reflection characteristics.

Recessed jambs

All windows have an indented lintel, sill, and jambs, creating space for hanging the curtains while at the same time ensuring that when the curtains are drawn, they fall in a way that creates a reasonably dead air gap between the curtain and the glass, improving the insulation characteristics.

Evaporative wind tower

A multi-directional wind tower fitted with evaporative cooling pads tops the stairwell. Outside air, cooled by flowing past the pads kept wet by circulation pump, falls down this well and enters the rooms through permanent ventilation openings above the doors. These openings may be closed with shutters if required.

Solar chimneys

South-facing thin-walled and dark-coloured shafts assist air exhaust in the summer days. The shafts are topped with fibreglass chimneys. Internal shutters cut out this exhaust when required (in the winter).

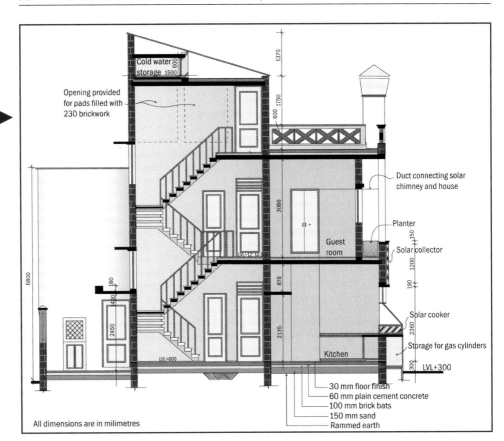

Building section highlighting integration of solar passive and active systems with the building envelope

Renewable energy systems

Architecturally integrated solar hot water

A 200 litre-per-day hot water system is architecturally integrated with the house. A solar water heater panel is mounted vertically on the south wall above the kitchen. This non-standard mounting angle reduces the problems of dust accumulation, endemic to Delhi. It reduces chances of breakage due to hail or cats, etc. It also frees up terrace space and is a technique that could be used in multi-storeyed buildings to provide individual hot water systems without using the roof. A rooftop insulated hot water tank collects the solar heated water, which is then available to all the toilets and kitchen. A sensitive thermostat activates the back-up low-wattage electrical element on the evening of cloudy days, but only if manually activated by the residents. An independent 100 litre-per-day vertically mounted solar hot water system is provided for the outhouse.

Built-in solar cooking

A sliding solar cooker is built in to the south wall of the kitchen. The pans can be accessed from inside the kitchen and come out at counter level for ease of use. Balance cooking is done with conventional fuel.

Waste recycling and water conservation

Compost pit

Two pits in the garden can ensure the composting of kitchen wastes. The kitchen, in turn, is provided with ample space below the counters to ensure separation of paper, organics, and other waste and recycling of the first two types.

Low water flushes

The house is fitted with flush valves and without any cisterns.

Energy-efficient lighting

Daylight

All spaces in the house are properly lighted with natural light in the day.

Low energy lights

All lamps and luminaries are based on warm colour CFLs (compact fluorescent lamps) so as to reduce electrical loads to the maximum.

Overall rationalization of loads

Including fans, lights, refrigerator, and a water circulation pump for the cooling tower, the running loads of the house work out to about 400 W. Adding a TV/computer or water boosting pump (not running simultaneously), the running load of the house is about 800 W. If one heavy appliance (electric iron, washing machine, etc., not running simultaneously) is added, the house can be run at under 2000 W peak all the time. There is no need to use electricity for space heating and cooling, nor for water heating. Using a careful selection of devices, 15 A circuits can be eliminated from the house altogether.

Inverter and future provision of photovoltaic charging

The house has a central inverter and a battery bank that can provide for 8 hours of emergency usage. It is planned that as soon as affordable, the house shall be put on photovoltaic charging and progressively made less dependent on the grid.

Performance

The thermal systems generally worked very well in winter, eliminating the requirement of heavy quilts altogether. The dry summer season performance was good except for slight droplets of water spilling over from the wet pads of the cooling tower into the staircase area. Monsoon comfort is dependent on strategically closing or opening the windows.

Renewable energy systems have worked well. Electrical systems are satisfactory, but highly stressed due to erratic power supply and extreme voltage fluctuations in the area.

Project details	Design features
Project description Plotted house for a couple *Architect* Sanjay Prakash *Climate* Composite *Consultants* Atam Kumar (energy) Sunil Arora (structure) *Project period* 1995–1996 *Size* 160 m² covered area in a plot of 450 m² *Client/Owner* Atam Kumar *Promoter* Cdr. C P Sharma (Retd) *Builder/Contractor* Asian Townsville Farms Ltd	▪ The house has been oriented to face south so that every habitable room has a liberal solar exposure. Shading has been carefully designed to prevent solar gains during summer and allow the same during winter ▪ External wall and roof insulation help the thermal mass to act in tandem with the inner space ▪ Reflective finishes of wall and roof for heat reflection ▪ Multi-directional evaporative wind tower atop stairwell, for cooling of house during dry summer ▪ South-facing solar chimneys assist air exhaust during summer ▪ Architecturally integrated solar water heating system and solar cooker ▪ Compost pits in garden for composting of kitchen wastes ▪ Low water flush valves without cisterns ▪ All rooms adequately daylit and artificial lighting is provided with low energy lights ▪ Overall rationalization of loads done by careful device selection ▪ An inverter and battery bank back-up has been provided with provision for photovoltaic charging in future

Temperature check for Delhi

(see Appendix IV)

Cold	Cool	Comfortable	Warm	Hot	
	•				January
	•				February
		•			March
			•		April
				•	May
				•	June
			•		July
			•		August
			•		September
		•			October
		•			November
	•				December

Add the checks

	3	3	4	2

Multiply by 8.33% for % of year

Heating	25
Comfortable	25
Cooling	50

Residence for Neelam and Ashok Saigal, Gurgaon

Editor's remarks

This residential unit located in Gurgaon uses simple passive architectural interventions to achieve thermal comfort in all seasons and uniform daylight. The architect has translated the traditional courtyard to a 'tropical skylight'—an innovative design response to composite climate. This concept performs all functions of a traditional courtyard and yet provides complete protection from rain and insects.

Architect Sanjay Prakash

A small residence in the composite climate of Gurgaon uses cost-effective passive techniques for achieving thermal and visual comfort

This house seeks to translate the traditional courtyard house of inner cities to an updated lifestyle of today. While remaining within the legal framework, which is based on the model of the semi-detached house for plots such as these, the design seeks to convert the small patio often given for light into a proud centrepiece of the house, a status that it once commanded in the traditional house.

Front view of the Saigal house: bedrooms have adjoining garden courts

The house is a ground-floor unit with a living/dining room, kitchen, and three bedrooms. The central dining area is built like a roofed courtyard and can allow the next floor to be made in the future without compromising the privacy or working of the space below.

Materials, techniques, and methods

Planning of built volumes and south orientation

The footprint of the house was shifted as far north as possible. The two important bedrooms were placed towards the front, south side, with the largest possible open area on the south, even though it would have been normally frowned upon since it is usually expected that the living rooms get an attached garden.

Courtyard and covered courtyard

The design pattern for this house is based on a pattern of nine squares. Taking inspiration from the traditional courtyard, the central space was

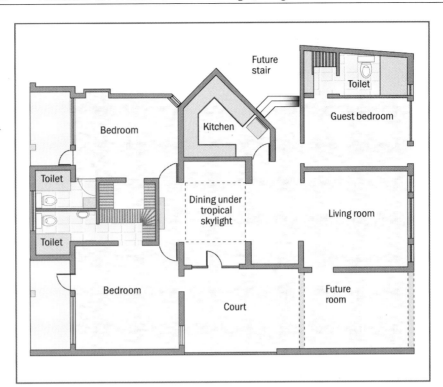

The plan of the Saigal house. Introvert planning with spaces organized around the central atrium, covered with a tropical skylight

designated the 'courtyard' despite the paradox that the house is a semi-detached quasi-bungalow. The plan that finally developed, in fact, contained two courtyards: the central (covered with the 'tropical skylight', see plan) court forming a permanent dining area and the next area a true temperate-weather patio.

The dining area with the tropical skylight

Daylight

The house is entered past the controlling kitchen window and almost from the rear due to the planning of built volumes. It is entered from a multi-faceted space where the future staircase can be. No rooms require artificial light even during the cloudiest of periods. The focus of the living areas is the central dining space, which ordinarily would have been the darkest room of the house, but is converted by the use of the 'tropical skylight' into the brightest space.

Tropical skylight

The central dining space was notionally thought of as a courtyard. However, neither insects nor rain nor modern lifestyle would have allowed this feature to be literally used from traditional models. Therefore, it was roofed over with what has come to be called the 'tropical skylight'. This is a skylight suitable for atria in tropical climates. It consists of the glazing being placed vertically, facing north and south, and covering the roof so that direct summer solar gain is avoided while winter sun streams into the space. The proper design of overhangs and glazing allows this geometry to be adapted to all tropical climates, and can be recommended instead of the horizontally glazed skylights which are an attractive but an unsuitable design feature for our climate.

Central evaporative cooling

Along the rim of the 'tropical skylight' are plenums to reach cooled air from an adapted desert cooler to every room in the house. It is an extremely simple affair, with two evaporating pads, a downward facing fan, and a pump.

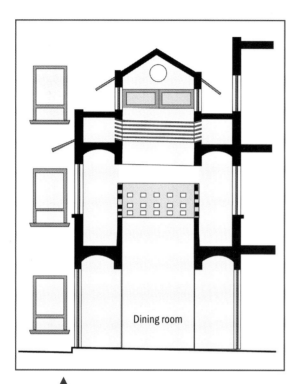

Section through the tropical skylight showing the plenum space for evaporative cooling system. The advantage of the skylight is that the construction of first floor would not deter daylighting

Fixed meshes

Fixed netting was installed inside the window and the glass windows opened outwards, operated by a winch pulley-type arrangement. The hardware failed miserably, but potentially is a real space and cost saver.

Reflective roof and insulation

The roof is reflective white (and mixed colour) broken China mosaic over 50-mm thick expanded polystyrene insulation. This makes the otherwise meagre cooling system very effective.

Optimization of structure

In order to achieve cost savings as well as reduce embodied energy, the concentration was on the structural design of roof slabs. The spans were limited to about 4 m, and the planning allowed a layout without a single beam or column. In fact, the house could have been roofed over with stone slabs (as has been done for the window overhangs). The total amount of reinforcing steel was reduced to as little as under 9 kg/m² (less than half of usual) while the reinforced concrete volume was restricted to under 0.12 m³/m² total.

Performance The thermal systems generally work well in winter, eliminating the requirement of all artificial heating, barring the occasional use of the fireplace. The performance during dry summer season is good. Monsoon comfort is dependent on strategically closing or opening the windows, but is still sufficiently uncomfortable so that a window air-conditioner was placed in the master bedroom for this season, with the children and parents using the room together at that time. Central cooling system is satisfactory, but requires maintenance a bit more often than usual window coolers.

Economics The initial cost of the project was the same as conventional. In fact, the use of self-managed construction and optimization of structural design and spans led to cost savings so the house was made for about 10% lower than the prevailing rates in that area at that time. The running cost is now in the form of electricity for the cooler (mostly) and air-conditioner (sometimes), and liquefied petroleum gas in the kitchen.

At a glance

Project details

Project description Plotted house for a couple with three children
Climate Composite
Architect Sanjay Prakash
Consultants In-house
Project period 1994
Size 110 m² covered area in a plot of 450 m²
Client/owner Ashok Saigal
Builder/contractor Owner-managed construction

Design features

- Southern orientation of important bedrooms
- Courtyard planning as a climatic response
- Central courtyard covered by 'tropical skylight'
- Centralized evaporative cooling for dry summers
- Roof insulation
- Optimization of structure to reduce embodied energy

Temperature check for Gurgaon

(see Appendix IV)

Cold	Cool	Comfortable	Warm	Hot	
	•				January
	•				February
		•			March
			•		April
				•	May
				•	June
			•		July
			•		August
			•		September
		•			October
		•			November
	•				December

Add the checks

	3	3	4	2

Multiply by 8.33% for % of year

Heating	25
Cooling	50
Comfortable	25

Dilwara Bagh, Country House for Reena and Ravi Nath, Gurgaon

Editor's remarks

This country house in the composite climate of Gurgaon is a good example of climate-responsive architecture designed and constructed using traditional materials and methods of construction. The adoption of passive architectural principles completely nullifies the use of energy-intensive space-conditioning techniques.

Architects Gernot Minke and Sanjay Prakash

A country house in the vicinity of Delhi uses traditional Indian architectural principles and methods of construction to provide updated requirements of an international lifestyle

The Dilwara Bagh (the country house) merges the traditional Indian courtyard house with the western 'prairie house' to meet the updated requirements of an international lifestyle. The house is a ground-floor unit with a living room, a dining room, kitchen, and three bedrooms. There are ample verandahs and gazebos. The dining area is a small courtyard with pool. There is an outhouse for guests and garages, four huts for servants and services.

▲ A south view of the house with windows for winter gains. The roof overhangs provide shading for overcoming summer gains

Materials, techniques, and methods

The lake

The flat agricultural land was transformed into an undulating garden. The earth for edge berms and other mounds was taken by forming a lake. This lake acts as a central visual element of the landscape design as well as a microclimate modifier. It stores the water of the monsoon and is topped up in dry season with a tube well, which was required in the first instance for irrigation. The lake is sealed with a clayey soil. The bank is designed and planted to allow the water level to fluctuate within a range of 0.6 m.

107 • Dilwara Bagh, Country House for Reena and Ravi Nath, Gurgaon

The vegetation

The drive up to the house takes the visitor through a playful series of views of the house and the lake contrasted with closed spaces with dense vegetation. A few other major trees accentuate some of the other shady areas where one can rest. The selection of the tree and shrub species is made keeping in view the practical requirement for obtaining a variety of fruits and getting natural compost.

The adobe in-fill walls with natural earthen colour gives the feel of a traditional village house

Planning of built volumes and south orientation

The single storeyed house is mostly set into the earth berms towards the north of the lake. The south side is exposed to the winter sun and shaded in the summer with overhangs and louvres.

Courtyard

The rooms are arranged around a central patio with a small pool with plants. This enables cross-ventilation for all rooms and cooling by evaporation.

Low energy structure

The plan was generated by a pattern of octagons and squares. The structural frame consists of load-bearing stone columns, which support beams and stone slabs to form slightly domical enclosures over all rooms, reminiscent of some of the temples in Dilwara in Rajasthan (hence the name of the house). The in-fill walls are from adobe blocks (handmade mud bricks).

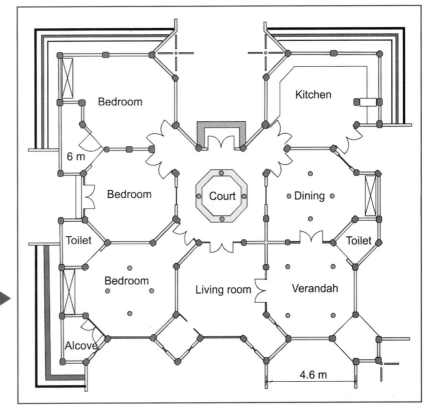

The building plan is a traditional courtyard house with a central courtyard flanked by rooms

◀ The building opens up to the south for winter gains

◀ Northern facade is partially earth bermed for optimum daylighting

South elevation

North elevation

Insulation and shading

A conical, light-coloured stone roof above the domical slab creates an air cavity and also provides reflection of solar radiation, besides shading the roof below. Wherever the berms cover the external face, an air cavity is formed by an inclined stone slab resting against the wall. All external surfaces of the building have either air cavity or summer shading by overhangs and louvres.

Daylight

The stone louvres are designed to take over the function of a usual steel security grille and at the same time provide sun shading and reflect the daylight into the rooms, acting like small light shelves.

Earth tunnel system

Additional cooling in the summer months is provided to each room by an earth tunnel system. The distance from the 2 kW fan to the building is about 60 m. The section consists of two masonry ducts at an average depth of 3 m below surface. The maximum air velocity is kept to 6 m.

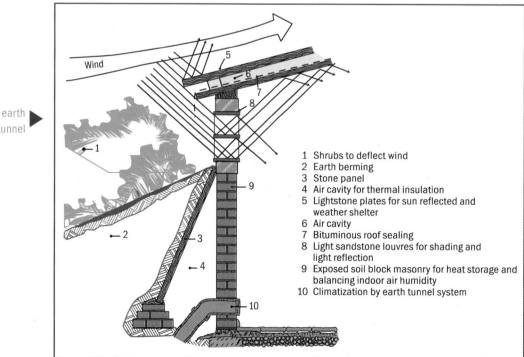

◀ Wall section showing earth berming and earth air tunnel

1 Shrubs to deflect wind
2 Earth berming
3 Stone panel
4 Air cavity for thermal insulation
5 Lightstone plates for sun reflected and weather shelter
6 Air cavity
7 Bituminous roof sealing
8 Light sandstone louvres for shading and light reflection
9 Exposed soil block masonry for heat storage and balancing indoor air humidity
10 Climatization by earth tunnel system

Performance The orientation works well in winter, and in combination with the tunnel system, eliminates the requirement of all artificial heating. The dry summer season performance is good. Monsoon comfort is dependent on strategically closing or opening the windows.

This earth tunnel system is probably the longest in personal use. The coefficient of performance has been studied to be as high as 20 at various seasons. The only problem reported is the absence of sufficient air circulation sometimes. This is because ceiling fans were not installed in order to appreciate the domical ceilings better. Floor-mounted fans are not very effective.

Economics The initial cost of the project was high, though not necessarily due to the essential features but due to the complicated sourcing of stone and craftsmen. The tunnel system eliminated the need for central air-conditioning (which could have been afforded by the owners) by accepting a system with lesser control. The running cost is now in the form of 2 kW of electricity for the tunnel system and liquefied petroleum gas in the kitchen.

At a glance

Project details

Project description Country house for a couple with two children

Architects Gernot Minke and Sanjay Prakash

Climate Composite

Consultants In-house

Project period 1992–1996

Size 206 m² covered area in a plot of about 16 000 m²

Client/owner Reena and Ravi Nath

Builder/contractor Architect-cum-owner managed construction

Design features

Summer
- Reduction of heat gain by
  - air cavity in walls and roofs
  - earth berms
  - shading by overhangs and louvres
  - shading by vegetation (trees and creepers)
- Increase of heat loss by
  - cross-ventilation
  - cooling through evaporation by water surfaces and plants (except during monsoon)
  - cooling through earth tunnel system

Winter
- Increase of heat gain by
  - direct gain through windows
  - underground earth tunnel
- Reduction of heat loss by
  - air cavities
  - compact building form

All seasons
- Balancing of temperature through thermal mass of walls and floors
- Balancing of indoor air humidity by earth walls (adobe)
- Increase of daylight by reflecting stone louvres in all windows
- Balancing of microclimate through water and vegetation

Temperature check for Gurgaon

(see Appendix IV)

Cold	Cool	Comfortable	Warm	Hot	
	●				January
	●				February
		●			March
			●		April
				●	May
				●	June
			●		July
			●		August
			●		September
		●			October
		●			November
	●				December

Add the checks

	Cool	Comfortable	Warm	Hot
	3	3	4	2

Multiply by 8.33% for % of year

Heating	25
Comfortable	25
Cooling	50

RETREAT: Resource Efficient TERI Retreat for Environmental Awareness and Training, Gurgaon

Editor's remarks

TERI has taken a major step by creating a model sustainable building complex, called RETREAT. This building would not put any pressure on the earth's fragile ecosystems and, in fact, actually regenerate what has been lost through neglect or misuse of nature's bounty. The building offers high level of visual and thermal comfort with a minimal environmental footprint and is replicable in part or in whole.

Architects Sanjay Prakash and TERI

A powerful and effective combination of modern science and traditional knowledge

RETREAT (Resource Efficient TERI Retreat for Environmental Awareness and Training) is TERI's vision realized—the vision of building a sustainable habitat, which is not just the first of its kind in this part of the world, but also one that inspires many such habitats to be created in the future. Springing from TERI's deep-rooted commitment to every aspect of sustainable development, RETREAT demonstrates effectively what a clear vision, sincerity of purpose, and sustained effort can accomplish.

Rising Phoenix-like from a swampy, degraded wasteland, the 36-hectare TERI campus at Gual Pahari, Gurgaon, today is a lush green, living and throbbing habitat, where RETREAT holds a place of pride. RETREAT is a part of TERI's Gual Pahari campus, about 30 km south of Delhi, in the northern state of Haryana. Built as a model training complex, RETREAT demonstrates efficient utilization of energy, sustainable, and integrated use of both natural resources and clean and renewable energy technologies, and efficient waste management.

Constructed over three years, and with a built-up area of 3000 m², this 30-room training hostel boasts of conference facilities for 100 people, dining space and kitchen, recreational area, computer room, and a library. What makes RETREAT unique is its total independence of the city's grid system and near-complete freedom from city services and infrastructure. Interestingly, energy planning in the building has led to a reduced load of 96 kW (peak) from a conventional 280 kW (peak), showing a saving of 184 kW (peak).

RETREAT is a combination of technology and architecture in a manner that the beholder's understanding of the environment, energy, and buildings deepens. An integrated design element, the solar passive architecture, renewable sources of energy, conventional and non-conventional heating and

A north view of the RETREAT
▼

cooling and energy-saving features—all stand out in an astonishing blend of the traditional and the modern.

Basically, three important things were considered in the creation of the complex. Firstly, the functionality of the building, and trying to see how energy is used in it. Secondly, the design of the complex minimizes demands of energy in the building by architectural intervention through passive concepts like solar orientation, latticework for shading, insulation, and landscaping. Thirdly, the space-conditioning and lighting demands are met through energy-efficient systems whereas the electric energy demands are fulfilled by renewable energy sources.

Passive designing for load reduction

Various passive design concepts have resulted in reduction of space conditioning loads by 10%–15%. Building envelope efficiency, which result in lowering of space-conditioning loads, was achieved by adoption of various passive techniques as listed below.

- The roof is insulated by using vermiculite concrete and China mosaic white finish. Walls are insulated by using expanded polystyrene insulation.
- Part of the building is sunken into ground in order to take advantage of ground storage and thus stabilize internal temperature.
- Shading devices and fenestration have been adequately designed to cut off summer sun and to let in winter sun.
- Glare-free daylight has been adequately provided in the conference hall, library, and recreation hall through use of specially designed skylights.
- Landscaping has been adequately designed so that wind directions are favourably altered. Deciduous trees are used in the southern side of the building to shade the building during summer. During winter, the trees would shed their leaves thus letting in winter sun.
- The building is oriented along the east–west axis so as to have maximum exposure along north and south. Architecturally, the building is consciously freed from the confines of a strict orientation in order to demonstrate that though energy-conscious architecture needs to be somewhat oriented, the orientation need not be rigid and interesting patterns can be formulated for architectural purposes. In RETREAT, the north block is made slightly concave towards the front, while the south block forms a hybrid convex surface facing the winter sun. The points of the south block broadly fall on the surface of large imaginary cones that generated the slightly free geometry and this allows the architecture to break away from the grid iron approach that is associated with 'solar architecture' normally.

Ensuring a sustainable supply of energy

TERI's Gual Pahari campus is intended to serve as a model of sustainable habitat based on new and clean technologies. Therefore, it makes full use of the most abundant

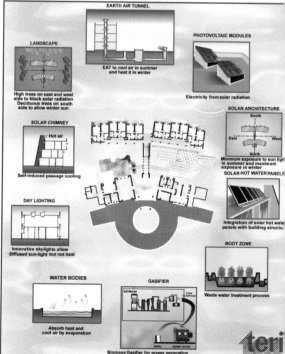

▲ Schematic diagram showing use of passive solar architectural principles, use of renewable energy sources and waste and water management in RETREAT

source of energy, namely the sun, by tapping the sun's energy in different ways, both directly and indirectly. Some of the innovative ways of tapping solar energy and using energy more efficiently at RETREAT are (1) solar water heater, (2) PV (photovoltaic) panels, (3) gasifier, (4) underground earth tunnels, and (5) waste water recycling.

Solar water heater

An array of 24 solar water heaters forms a part of the parapet wall of the living quarters. The system can deliver up to 2000 litres of hot water (at 65 °C) every day. In winter, when the days are short and the sun less intense, gas derived from burning twigs, dry leaves, etc. serves as a back-up source to heat the water. The heat given off when the stand-by generator is running is also collected and utilized in the solar water heater, as a back-up source.

▲ South view of the building showing solar water heating panels and solar chimney

◄ The sun is the powering force of RETREAT, where solar panels are used to form a 'solar roof'

Photovoltaic-gasifier hybrid power plant

The building is powered by a PV-gasifier hybrid system. Firewood, dried leaves, twigs, and crop residues after the harvest and such other forms of biomass fuel the 50 kW gasifier. The 10.7-kW (peak) roof-integrated PV system generates power from solar energy. The power available from both the gasifier and the systems is managed and controlled with the help of a building management system. The excess power generated from the sources goes to charge the battery bank. During the night, the loads are met by the 900 amp-hours/240 V battery bank. The external lights and water pumps are powered by independent stand-alone PV systems.

Most of the external lights located outside the building are powered by independent panels. Each light has its own pair of small PV panels (roughly a metre wide and half a metre tall) and is thus a self-sufficient 'stand-alone' unit. The water pump is also powered by a PV panel.

▲ The biomass gasifier is the main source of power during the day

Energy-efficient systems

The underground earth tunnels

The living quarters (the south block) are maintained at comfortable temperatures (approximately between 20 °C and 30 °C) round the year by circulating naturally conditioned air using earth air tunnel system, supplemented with a system of absorption chillers powered by LPG (liquefied petroleum gas) in humid season and air-washer in dry summers. Underground structures are not exposed to the sun and thus do not heat up as much. Secondly, the surrounding earth insulates them, which helps in maintaining a more or less constant temperature. Temperatures recorded at roughly 4 m below the surface show that they are stable and reflect the average annual temperature of a place.

However, the cooler air underground needs to be circulated in the living space. Each room in the south block has a 'solar chimney'; warm air rises and escapes through the chimney, which creates an air current: the cooler air from the underground tunnels rushes in to replace the warm air. Two blowers installed in the tunnels speed up the process. The same mechanism supplies warm air from the tunnel during winter.

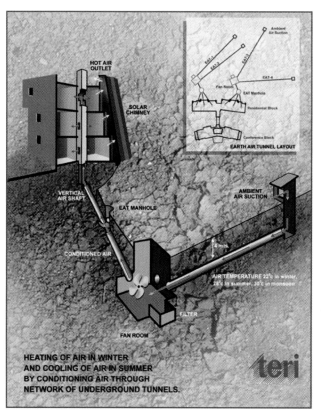

◀ Passive space-conditioning using earth air tunnel system

Absorption chillers

A set of eco-friendly chillers, which run on LPG and require minimum electricity, provide extra cooling when needed. LPG being a non-renewable source of energy, efforts are under way to run the chillers on producer gas generated by the wood-based gasifiers.

The conference centre, which accommodates up to 100 participants, is conditioned by means of the ammonia-based absorption chilling.

Energy-efficient lighting

RETREAT uses the energy-efficient compact fluorescent lamps in residential quarters, corridors, lobby, and toilets. Energy-efficient tubelights with electronic chokes are used for conference halls, recreation room, computer room, dining hall and in administration areas. The conference rooms enjoy glare-free daylight through strategically placed skylights.

Time-based controls switch off lights at preset time. Key tag system is installed in the rooms for energy conservation.

Waste management

Waste water recycling by root zone system

Waste water from the RETREAT is recycled using the root zone technique. It is a natural waste water treatment process based on aerobic and anaerobic decomposition of the contents in the roots of the reeds (phragmytes) and microbial organisms. The process is natural, economical, and efficient and gives quality treated water. This water is used for irrigation. The entire area is proposed to have water harvesting and watershed management.

Monitoring the system

The RETREAT building is an outcome of TERI's relentless research in sustainable building design. It is also an evolving experiment: the information gathered on how well the building really performs under varying ambient conditions – in winter and in summer, on bright days and on cloudy days, at varying levels of occupancy, and so on – will help design better systems, at Gual Pahari and elsewhere. The elaborate, extensive, and sensitive network of sensors, linked to a central station, monitors every heartbeat of the system 24 hours a day, 365 days a year.

The extensive building monitoring system logs in all relevant temperature, humidity, and power data. Thus, the facility is also a first-rate source of immense quantities of scientific data that can power many more experiments and indeed influence the design of other such facilities the world over.

▲ Water and waste management system

Performance The winter temperature in the rooms heated by solar gains and earth air tunnel systems was recorded to be 22 °C (average night-time condition) when the ambient temperature was about 10 °C. In the dry summer month of May, room temperature (in the rooms cooled by earth air tunnels combined with evaporative cooling system) of 28 °C (average daytime condition) with 45%–50% relative humidity was recorded when the ambient temperature and relative humidity were 40 °C and 30% respectively. In the humid summer months, room temperature (in rooms cooled by earth air tunnel supplemented by ammonia absorption chillers) of 30 °C and 65% relative humidity were recorded with the ambient condition being 38 °C at 70% relative humidity. The conference rooms cooled by ammonia absorption chillers maintain an average temperature of 25 °C at 55% relative humidity. The building being only partially loaded as yet now consumes a maximum of 40 units of electricity per hour. The average units generated by the PV system is 55 units per day on a sunny day.

Conclusion A great deal of thought and planning have gone into the construction of RETREAT, but it is not just a facility: it is a concrete reaffirmation of TERI's faith in its research and of its commitment to sustainable development—it is TERI's gift to humankind in the 21st century.

At a glance

Project details

Building/project name RETREAT (Resource Efficient TERI Retreat for Environmental Awareness and Training)
Site address Gual Pahari, Gurgaon
Building type Institutional
Architects Sanjay Prakash and TERI
Contractors and system suppliers Confoss Constructions Pvt. Ltd (Civil); Janus Engineering Pvt. Ltd (Electrical); Pyrotech Electronics Pvt. Ltd (BMS); Suvidha Engineers India Ltd (HVAC); Shiva Furnishers (Interiors); Thermax (HVAC and Root zone system); Tata BP Solar (Solar Photovoltaics); Jain Irrigation (Solar hot water)
Year of start/completion 1997–2000
Client/owner Tata Energy Research Institute, New Delhi
Covered area 3000 m²
Cost of the project Civil works - Rs 23.6 million; Electrical works - Rs 2.5 million
Cost of various technologies Rs 18.54 million

Break-up

■ Earth air tunnel	Rs 1.68 million
■ Solar chimney	Rs 0.28 million
■ 10.7 kW Solar photovoltaic system	Rs 7.40 million
■ Stand-alone PV streetlighting system	Rs 0.28 million
■ Root zone system	Rs 0.85 million
■ Solar water heating system	Rs 0.25 million
■ Building management system	Rs 2.30 million
■ Ammonia absorption cooling system along with gas bank	Rs 3.10 million
■ Air-conditioning system (HVAC)	Rs 2.40 million
Total (cost of technologies)	**Rs 18.54 million**

Design features

Orientation, insulation, and design of the building

- Wall insulation with 40-mm thick expanded polystyrene and roof insulation using vermiculite concrete (vermiculite, a porous material, is mixed with concrete to form a homogenous mix) topped with China mosaic for heat reflection.
- Building oriented to face south for winter gains; summer gains offset using deciduous trees and shading.
- South side partially sunk into the ground to reduce heat gains and losses.
- East and west walls devoid of openings and are shaded.

Earth air tunnel for the south block

- Four tunnels of 70-m length and 70-cm diameter each laid at a depth of 4 m below the ground to supply conditioned air to the rooms.
- At a depth of 4 m below ground, temperature remains 26 °C (in Gurgaon) throughout the year.
- Four fans of 2 hp each force the air in and solar chimneys force the air out of rooms.
- Assisted cooling by air washer in dry summer and a 10 TR dehumidifier in monsoon.

Ammonia absorption chillers for the north block

- Gas-based system with minimal electrical requirement (maximum 9 kW).
- Chloroflurocarbon-free refrigerant (ammonia).

PV-gasifier hybrid system

- 50 kW gasifier and 10.7 kWp solar photovoltaic
- Generates producer gas (containing methane) which runs a diesel generating set with 70% diesel replacement.
- 1 unit of electricity produced needs 1 kg of biomass and 90 ml of diesel.
- 900 amp-hours batteries at 240 V.
- 36 kVA bi-directional inverter.
- Load manager controls and manages loads

Solar hot water system

- 24 solar water heating panels (inclined at 70 degrees instead of 45 degrees) integrated with parapet wall.

Lighting

- Lighting load 9 kW (reduced from a minimum of 28 kW in a conventional building).
- Lighting provided by compact fluorescent lamp, high efficiency fluorescent tubes with electronic chokes.
- Lighting controls to reduce consumption (timers, key-tag systems).
- Innovative daylighting by means of skylights.

Waste water management system by root zone system

- Cleans waste water (5 m^3/day) from toilets, kitchen, etc.
- A bed of reed plants (phragmytes) treat the water and the output is used for irrigation.
- The plants take up nutrients from the water and thrive on the same, in the process cleaning the water.

Building management system

- Monitors building parameters (temperatures, humidity, consumption, etc.)
- Monitors electricity generated from each source
- Decides on load-sharing and load-shedding to optimize energy usage
- Records at regular intervals

Temperature check for Gurgaon

(see Appendix IV)

Cold	Cool	Comfortable	Warm	Hot	
	●				January
	●				February
		●			March
			●		April
				●	May
				●	June
			●		July
			●		August
			●		September
		●			October
		●			November
	●				December

Add the checks

	3	3	4	2

Multiply by 8.33% for % of year

Heating	25
Comfortable	25
Cooling	50

Water and Land Management Institute, Bhopal

Editor's remarks

The low-rise institutional building, which merges with the landform, has used simple yet innovative passive solar techniques to reduce space-conditioning loads. Uniformly daylit round the year and with favourable orientation, building form, and detailing, this building is a unique example of bioclimatic architecture.

Architect Sen Kapadia

This landscraper is being built on the conviction of providing maximum human comfort, without relying too much on conventional energy consuming system

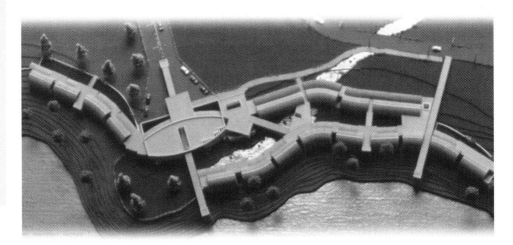

 A view of the WALMI building. The flowing form overlooks a water body which has been used to advantage for modification of the micro climate

Introduction

WALMI (Water and Land Management Institute) was established in Bhopal, Madhya Pradesh, under the central government aid, with a desire to increase the effectiveness of irrigation projects. The Institute conducts research and training programmes both in built facilities and on fields.

Site

The site is located on a flat top hillock adjoining the Kaliasote dam just 15 km away from the new market. The site offers an undisturbed view of the water body, down the northern slopes and experimental farms, down southern slopes of this 30-m high hillock.

Design, materials, and techniques

WALMI is viewed largely as an environmental planning apparatus. The Institute functions in a low profile but highly nature-sensitive atmosphere. The life here is self-sufficient in work, recreation, living and socio-cultural activities amidst lush landscape that at once provides natural beauty and microclimatic control.

Planning and built form

The 230-m-long institutional building of WALMI is planned as a 'landscraper'. If built vertically, it would be a 60-storey 'skyscraper'. The building exploits the north–south orientation for user comfort. The WALMI

▲ Part plan of WALMI as a 'landscraper'. The building is oriented along east–west axis with maximum exposure on the north and south

1 Classroom
2 Bridge
3 Cafe lounge
4 Toilets
5 Hydrology lab
6 Computer room
7 Library
8 Courtyard
9 Snack bar
10 Audio-visual room
11 Auditorium lobby
12 Auditorium
13 Faculty room
14 Store

administration and hostel buildings follow the contours of the hill on which they sit, while the long horizontal plan ensures a rural character and simple construction. This also affords all users to enjoy the landscape at a personal level as opposed to gardens seen from a higher floor as a visual mass.

The long east–west axis determines an axial plan. As can be seen from the plan, this provides an advantage of giving strategic location for main building; secondary location for hostels; and quiet, secluded area for residential units, with area for expansion.

This master plan is integrated with landscaping plan to provide a 'habitable landscape' rather than buildings, roads, and gardens. This is a new heightened blend of architecture, nature, and people in cooperative existence. A large part of the site has been developed as 'green zones' of indigenous trees that require minimal maintenance. The plan envisages an integrated development of indoor and outdoor spaces in total harmony with the local landscape.

Usage of natural building materials and indigenous plants ensures appropriate aesthetics.

Landscaping

Trees are planted primarily to be effective for cooling, reducing the dust, sun glare, and acting as natural shields against hot and cold winds. Trees like in ancient times provide inspirational contemplation spots.

Water as a microclimate modifier

Water has also been a central and sacred feature of human settlement since Vedic period. The presence of water in the form of stream, cascades, and pools is thus a significant element of the WALMI plan. Planning carefully avoids European and Moghul traditions of vast sheets of formal reflective stagnant water bodies. Instead, a moving water body is at once alive and free from mosquitoes. Additionally, water within the campus make a meaningful visual and physical dialogue with the reservoir of Kaliasote dam.

Water is pumped up from Kaliasote reservoir and stored in overhead water tanks located on top of buildings. Cascading water is extensively used to

A typical section showing passive solar features of WALMI buildings

1 Ground cover
2 Water sprinkler
3 Insulated roof
4 Shading trees
5 Water trough

provide coolant to air-conditioning plants, and for sprinklers provided on rooftops. Eventually this water is recycled to landscaped zones. All this has been detailed as a showcase for demonstration of a natural element in true harness and as an appropriate modifier of the microclimate. This along with land formations effectively suggests creative interpretation of WALMI.

The essential element of this project is the modular unit. The 25 × 7 metre unit is developed for easy adaptability for research laboratories, library, and administrative functions as it constitutes the main bulk of the built form. As the Institute progresses, modular units will combine to generate a gently curvilinear building that responds to the natural contours of the hillock and a spectacular view of the lake.

Circulation

Pedestrian circulation is given prime importance in this plan. Meandering pathways through different spaces aligned with flowering and fragrant shrubbery provide pleasant atmosphere while commuting to different zones.

Ventilation, daylighting, and roof treatment

Deep and single-loaded corridor on the south ensures cross-ventilation, and avoids heat and glare in all rooms. Window openings are oriented on the north side to allow abundant daylighting of all rooms.

Using natural materials, most areas of buildings are built with standardized elements such as lintels and storage units. Flooring employs local natural stones.

The building, largely located on the east–west axis, enjoys extensive north exposure. This allows induced cross-ventilation and abundant natural light without any heat gain, as west side exposure is minimized and the roof is also insulated. This helps energy savings by reducing electrical consumption for lights and fans. Rooftops are fitted with water sprinklers that cool through evaporation. The roof is also insulated with 2.5-cm layer of thermocole over RCC (reinforced cement concrete) shell and topped with stone tiles. Appropriate trees are placed judiciously to cut the glare and heat gain, while built-in desert coolers ensure pleasant environment at a reduced energy use.

The sprinkler system is integrated with the roof. This reduces heat gain by evaporative cooling during dry summers

All passive systems are self-induced solutions based on ongoing clinical research. The building has been designed to be cooler by 10 °C with reference to ambient external temperature. The aim is to provide effective human comfort and reduced load on air-conditioning.

Services

Electrical supply

Although electricity will be supplied at the site, investigating the use of photovoltaic based local generation of power is recommended. Buildings are planned to virtually eliminate electric lights during daytime.

Sewage management

There is no municipal disposal on site and this is being managed by the two *gobar* gas plants. These plants not only effectively manage waste disposal but also yield fertilizers and methane gas.

Performance

The architect does not have any performance data of the building. However, the architect feels that bioclimatic consideration of the building design along with thoughtful incorporation of passive measures would ensure thermal and visual comfort in the building round the year with minimal energy usage.

Conclusion

Taking advantage of a unique site, away from constraints of a city, a master plan of WALMI is created as a special but appropriate material embodiment of an institute's desire to harness bounties of nature.

Plan and architectural details vastly differ from conventional plans, only to economize the expenses and create harmony with nature. As an extension of the sponsor's – the EPCO (Environmental Planning and Co-ordination Organization) – desire to create exemplary architecture, the WALMI plan strives to answer physical needs with creative integration of planning, landscaping, and appropriate technology. When applied in totality, the resultant cost-effectiveness and its aesthetic value are a natural consequence in this passive solar architecture.

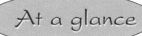

Project details

Project name Water and Land Management Institute, Bhopal
Building type Institutional
Architect Sen Kapadia
Year of start/completion 1983–87
Climate Composite
Built-up area 10 800 m² (when fully constructed)
Owner Water and Land Management Institute, Bhopal

Design features

- The building is oriented along the east–west axis to have north–south exposure. This reduces solar gain as well as glare. Simultaneously, there is maximum advantage of natural daylighting
- Use of high bulk local stone walling adds to thermal mass.
- Adjoining areas with soft ground cover and drip irrigation avoid all reflected heat and glare
- The roof is insulated with 2.5-cm layer of thermocole over RCC shell and topped with stone tiles. Additional water sprinklers on the roof allow cooling through evaporation
- On the lower floor, the air is admitted through a built-in trough of water ensuring higher humidity and coolth during dry hot summer months
- The surroundings are suffused with dense planting of evergreens and water bodies

Baptist Church, Chandigarh

Editor's remarks

This religious building, located in Chandigarh, has been designed in response to the climate. Several solar passive measures have been used to reduce its energy demand.

Architects Surinder Bahga and Yashinder Bahga

An environment-responsive design attitude specific to local culture opposed to western inspired hybrids

The Baptist Church, Chandigarh, represents an environment-responsive design attitude specific to the region and local culture and is opposed to the western inspired hybrids. The Church building has an identity of its own and stands out amidst box-type architecture of nearby housing colonies.

▲ A south-west view of the Baptist Church, Chandigarh

Building has been designed within a fixed volume of 11.3 × 32 × 11 m and several other strict by-laws of Chandigarh Administration. Entrance verandah, prayer hall for 250 persons, musical rehearsal hall, stores, and toilets are in the ground floor. The first floor houses a seating balcony for 100 persons, office, priest's residence, baptism tank, and changing room. The second floor comprises a fellowship hall, office, library, toilets, and a guest accommodation.

Design materials and methods of construction

Various measures were taken while designing the building to make it energy conserving. An integrated approach for both the economic and physical parameters resulted in evolution of building form, which by itself helped in minimizing the quantum of energy used in construction, maintenance, and performance.

Materials and finish

Heat gains are prevented by providing cavity walls in the north-east and south-west. Minimum use of glass, provision of cavity walls, and white-painted barrel vaults in roof, which reflect sun rays, contribute a lot in maintaining cool and comfortable interiors in summer. The external surface has been plastered with 2-cm thick layer of white chips mixed with white cement and ordinary portland cement for adding reflectivity to the roof.

Landscaping

A number of Ashoka (*Polyalthia congifolia*) trees on the site boundaries considerably modify the microclimate and provide shading.

Daylighting and cross-ventilation

Lighting and ventilation, if well planned, contribute to the quality of life by increasing comfort, safety, and performance. Fenestrations are planned to ensure good cross-ventilation which helped in reducing the load on cooling devices in summer. The cool air that enters from the lower windows becomes warm and exits from top, maintaining airflow. Windows are deeply recessed in the walls to provide adequate shading from direct sunlight. Natural light is provided to all spaces as far as possible. So no artificial lighting is required during the day. Interiors are plastered and painted white to get better reflectivity. Energy-efficient lighting equipment are provided to minimize energy consumption.

▲ Light-coloured finishes on the walls and roof, deep recessed windows are some of solar passive measures adopted in this Church building

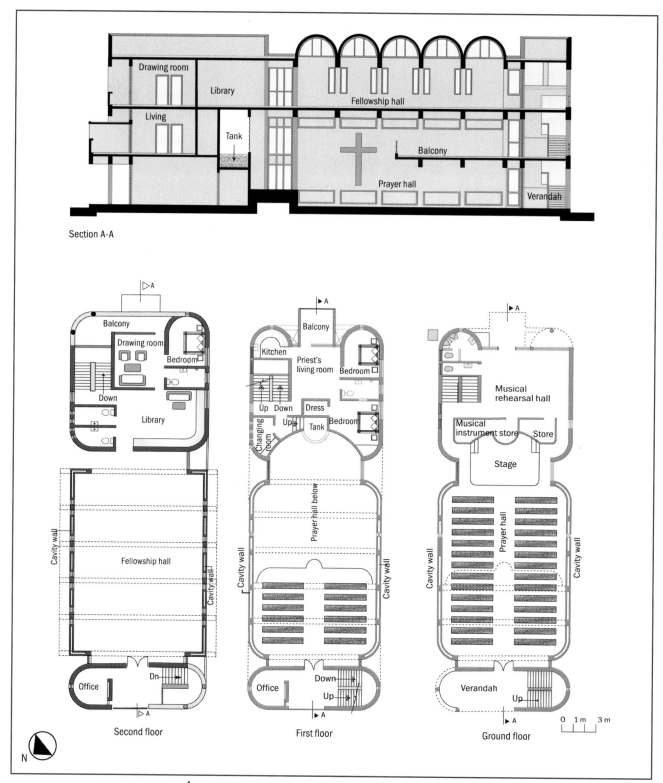

A diagram showing plan and sections of the building

Performance The building remains cool during summers and is able to provide thermal comfort condition without the need for additional air-conditioning. Adequate cross-ventilation is ensured and fans provide the desirable comfort conditions. The building is well lit round the day and eliminates use of artificial lighting during daytime.

Five per cent of the total project cost was incurred additionally in order to incorporate energy-efficiency measures.

At a glance

Project details

Building/project name Baptist Church
Site address Sector 44-C, Chandigarh
Building type Religious
Climate Composite
Architects Surinder Bahga and Yashinder Bahga
Year of start/completion 1987–1989
Client/owner North-West Indian Baptist Association, Chandigarh
Plot area 755 m²
Covered area 900 m²

Design features

- Cavity walls on the north-east and south-west minimize heat gains.
- Strategically located trees provide shading and modify micro climate.
- Deep recessed windows cut direct solar gains.
- Light-coloured finish of walls and roof reflects heat.
- Proper fenestration design ensures natural lighting and adequate cross-ventilation.
- Optimized use of glazed surfaces minimizes heat gains.
- Use of lighting equipment that are energy-efficient.

Solar Energy Centre, Gual Pahari, Gurgaon

Editor's remarks

The Solar Energy Centre uses several passive solar techniques to counter the composite climate of Gurgaon. However, due to maintenance problems, the roof evaporative cooling system, which was designed to provide thermal comfort during dry summer, is not being used in the technical and administration blocks and this leads to higher indoor temperature. Daylighting is adequate in all areas.

Architect Vinod Gupta

Demonstration of passive and active solar systems and use of innovative fenestration design to achieve thermal and visual comfort

High quantity of energy is consumed for providing lighting, ventilation, and thermal comfort in buildings. By proper planning and design, the architect has made it possible to reduce, and in some cases totally eliminate energy wastage in lighting and space conditioning. The buildings of the SEC (Solar Energy Centre) have been designed to make full use of on-site energy for providing environmental comfort. The complex is located in the composite climate of Gurgaon, about 35 km south of Delhi. The buildings in this complex has been divided into three groups.

- The technical and administration block comprising administration, library, cafeteria, laboratories and testing areas, and the solar simulator section.
- Workshop building and electric sub-station, etc.
- The guest house building.

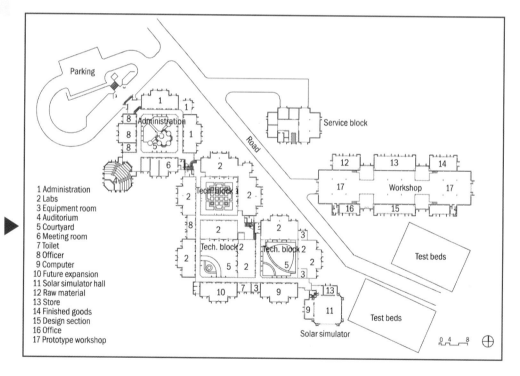

A site plan showing the layout of the administrative, technical, and workshop building

Technical and administration blocks: materials, methods, and techniques

Passive solar features

Roof and wall treatment
This double-storey research centre is located in a suburban setting measuring about 200 acres. The site constraints being minimal, the architect had freedom to determine an appropriate form and, therefore, the building was conceived as a low spread-out structure, arranged around courtyards,

A view of workshop building

maximizing the use of a roof surface evaporative cooling system with automatic controls. For comfort cooling, the strategy used was to reduce ingress of heat by using hollow, concrete block walls, properly shaded windows, and a reflective finish on the roof surface.

Windows for daylighting

Special, openable louvred shutters were designed for the east- and west-facing windows. Aluminium windows were used in preference to steel windows as this gives tighter fitting windows with lesser infiltration. The windows were split into two parts, one located at the normal height and the other just below the vault.

The lower windows provide ventilation and view and some daylighting close to the window. The upper windows ensure daylighting deep inside the work space even if partitioned. To provide daylighting in all work areas, and to ensure that artificial lights are not used during daytime, all deeper office rooms have been provided with windows on two different sides. This provides better distribution of daylight and eliminates contrasting illumination levels. The windows have been properly shaded to reduce glare from the sky.

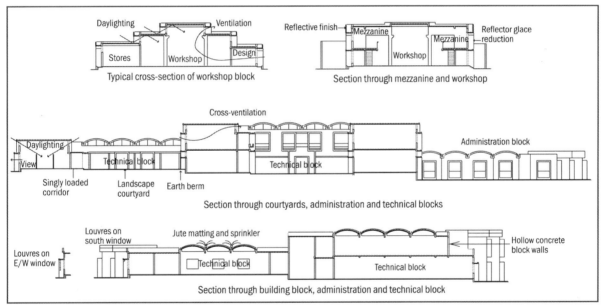

A section of the technical and administration blocks showing details of ventilation and daylighting and roof evaporative cooling system

This is a group of buildings primarily for day use. Most of the SEC staff work here and thermal comfort is hence an important issue. Some of these buildings also house special laboratory equipment and computers requiring humidity and temperature control. The design strategy for these buildings has been carefully chalked out.

For the winters, all rooms have been provided with windows so that they would receive sunlight for at least half the day. The monsoon season merits special consideration, as evaporative cooling does not work during this period. In these buildings, cross-ventilation has been provided in all the rooms.

Insulation
Mechanical air-conditioning has been provided for special equipment rooms only. The roofs of such rooms have been specially insulated to reduce the cooling load on the air-conditioning plant.

Artificial lighting
Fluorescent lights, which consume only a third of the energy consumed by ordinary incandescent lamps, have been provided in all the office areas. These would be used only in the evenings and on especially dark days.

Workshop building *Passive solar features*

Ventilation and daylighting
These buildings will house more machines and few humans. Since a great deal of heat would be generated within the building itself, the building cross-section is designed to facilitate removal of warm air. Permanent ventilators have been placed on the south side, the roof is insulated and two-level windows have been provided for good ventilation. To prevent heat gain through the roof, it has been given a reflective white finish. The entire building is daylit and no artificial lighting would be required during normal working hours. This building has an interesting system of daylighting. The building has a stepped cross-section with a reflective finish on the roof surfaces. Daylight is reflected from the lower roofs into the building providing glare-free even lighting. The drawing offices have deeply recessed windows with baffles, which cut out glare.

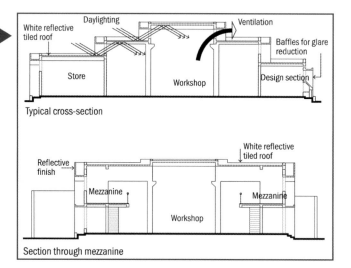

Section of the workshop building demonstrating use of natural lighting and ventilation

The workshop building is daylit using innovative window design

Guest house building

Although this was the first building to be completed in the campus of the SEC, it has been put to use sparingly.

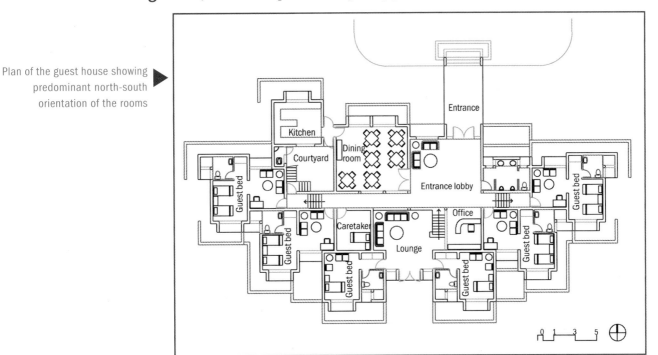

Plan of the guest house showing predominant north-south orientation of the rooms

Passive solar features

Earth berming

The south slope facing an artificial water body was chosen for this building. It has a white painted reflective finish on the exterior and is mainly oriented to 20 degrees east of south. Earth shelter is the main cooling system and the guest rooms are partially sunk into the ground. The roof has a terrace garden, which is watered during the summer months facilitating evaporative cooling. A special section is used to ensure cross-ventilation of each guest room.

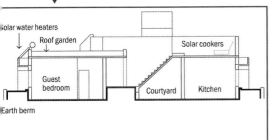
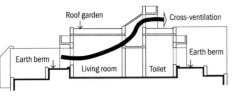
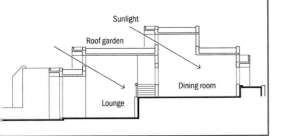

Section of the guest house highlighting earth-berming, roof garden, solar ingress, roof garden, and solar water heaters

Specially designed windows for winter heating

Windows are provided with an arched sunshade, which allows the winter sun to come in even through the top part of the glazing. The windows have been designed to reduce infiltration of outside air.

Solar water heating

Solar water heaters integrated with the architectural design have been provided for each toilet. They have been deliberately placed in a highly visible location so that their state of

A view of the Solar Energy Centre guest house, which is partially earth-bermed. The solar water heaters form an integral part of the building

disrepair (dust accumulation, leaks, and breakages) can be monitored easily. Back-up water heating is provided in the form of an outline electric heater located inside the toilet and not as a heating element built into the hot water tank. The latter is a waste of electrical energy. A sunny terrace for solar cooking has been provided near the kitchen.

This building will be used at all times of the day and comfort required is of a higher order. Because of the variety of activities in it, the energy conservation potential is the greatest in this building. Since this building would also be used at night, extremely efficient artificial lighting has been provided here. Bedrooms are equipped with small discharge lamps, while fluorescent lights have been used at all other places. No air-coolers or air-conditioners have been used here.

At a glance

Project details

Site 200 acres of land in Gurgaon
Climate Composite
Building types Institutional/residential
Architect Vinod Gupta
Building/project name Solar Energy Centre
Year of start/completion 1984–1990
Client/owner Solar Energy Centre, Government of India
Covered area 6943 m²
Cost of the project Rs 15.5 million (excluding cost of renewable energy technologies)

Design features

Technical and administrative block
- Courtyard planning with single-loaded corridors for ventilation and landscaped courtyard to modify microclimate
- Hollow concrete block walls to reduce heat gains
- Properly designed windows and shading devices
- Provision for rooftop evaporative cooling
- Insulation for air-conditioned blocks

Workshop building
- Building section developed for ventilation and daylighting
- Heat gain by the roof minimized by insulation and reflective roof finishes

Guest house
- Built on the south slope of an undulating site, and partially earth-bermed from three sides. Ground coupling is an effective means of improving the indoor thermal comfort.
- Terrace garden is watered during the summer months. The evaporation of water modifies the microclimate and also absorbs a major part of the cooling load in summer.
- A special section of the roof provided with manually-driven ventilators to ensure cross-ventilation of each guest suite.
- External surfaces of the building finished with white reflective paint to reduce the absorption of the incident solar radiation on the exposed portions of the vertical walls.
- Windows protected by arched sunshades (overhangs and sidewalls) of predetermined dimensions to avoid the entry of direct sunlight into the building during summer, but to allow it during the winters.
- Solar water heaters integrated with the architectural design have been provided for each bathroom.
- A sunny terrace provided near the kitchen to facilitate solar cooking.

Temperature check for Gurgaon

(see Appendix IV)

Cold	Cool	Comfortable	Warm	Hot	
	●				January
	●				February
		●			March
			●		April
				●	May
				●	June
			●		July
			●		August
			●		September
		●			October
		●			November
	●				December

Add the checks

	3	3	4	2

Multiply by 8.33% for % of year

Heating	25
Comfortable	25
Cooling	50

National Media Centre Co-operative Housing Scheme, Gurgaon

Architect Vinod Gupta

Aims to achieve resource conservation in a group housing society with varying requirements and finances

Editor's remarks

Climate-responsive built forms for a group housing society with 180 dwellings, each having varying requirements and finances. Innovative waste and water management techniques are employed to achieve sustainability.

Co-operative housing societies in Delhi suffer from many problems. They have to be designed within the limits imposed by the zoning regulations, which prescribe the total built area and the ground coverage. When these two are combined with fire regulations, they result in a very inflexible and uninteresting development. The alternative to group housing is plotted development, which starts and ends with a layout plan within which house owners build according to their needs and designs. The result is lack of architectural coherence and an excessive emphasis on individual buildings at the expense of overall environment. However, there are traditional examples of urban design that allow for individuality within an overall design framework like in the old city of Jaipur where buildings retain their identity without damaging the context.

A view of the group housing from the approach road

The housing scheme for the NMC (National Media Centre) at Gurgaon consists of 180 houses. The members are journalists, artists, medical practitioners, and teachers. The project began as a group housing scheme that was later converted to plotted housing, which created an unusual situation where members wanted all the advantages of a group housing scheme in the plotted development. All NMC members wanted complete houses and not just plots, even though their requirements and finances varied considerably. There were those who wanted a very small unit of 70 m² while others were not satisfied with 300 m². This called for a system of design and construction where individual requirements could be accommodated.

Each house is equipped with a solar chimney where a single evaporative cooler is fitted and by minimal ducting the entire house is cooled

Perhaps the most interesting part of the scheme is the separation between pedestrian and vehicular traffic. The ring road is designed for cars and car parking and the houses themselves are approached from cul-de-sac roads. This has resulted in a very safe environment for young children while allowing vehicular access to every house.

Features

Apart from the attempt at creating houses suited to the individual owner's needs, the project addresses several other problems.

The houses are designed to be thermally comfortable and energy efficient. Roofs are insulated with earthen pots laid in mud phuska. The buildings are oriented along the favourable north–south axis but with sufficient variation so as to avoid monotony. Windows are protected with appropriately designed sunshades and west-facing walls are shaded by trees. Every house has a sunny and a shaded side for winter and summer use.

Each house is equipped with a chimney where an evaporative cooler can be fitted with very little ducting; it is possible for the entire house to be cooled by the cooler. All houses have plumbing for fitting solar water heaters.

There is no municipal water supply in the area and all colonies depend upon tube wells for their water supply. The underground water table is threatened by overuse and poor recharge. Buildings and roads surfaces do not allow rainwater to percolate into the ground. At NMC the impervious hard paved areas have been minimized. Parking areas and service lanes are paved with open jointed bricks which

A system of soak pits has been provided in the central green area, which are, in turn, connected to the drainage system that allows the surface run-off to be put back into the ground

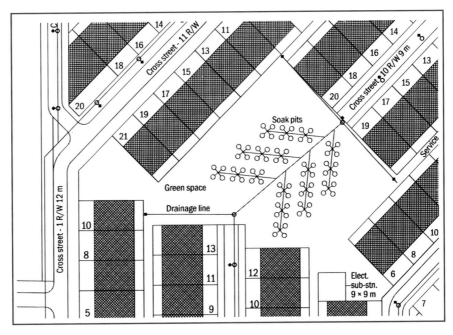

allow percolation. To compensate for the buildings and the roads, the percolation capacity of the green areas has been enhanced by providing a series of soak pits connected to the drainage system. This system of rainwater harvesting is so effective that most of the rainwater is absorbed within the site and even after a heavy downpour there is little outflow from the colony. It is hoped that this system will preserve the underground water supply of the colony.

In the absence of a system of municipal garbage collection, garbage disposal poses a challenge. A place has been provided for vermiculture where all the garbage from the colony is deposited. It is hoped that the residents will put into a place a system of sorting garbage into biodegradable and non-biodegradable materials within the house itself.

The power supply in the area is erratic and a standby source of power supply is needed. The wiring in each house has been done in a manner that makes the installation of an inverter easy.

Maintenance of open spaces in housing is a problem because these are public spaces, highly susceptible to encroachment. This is easily resolved in this scheme by planning them such that they remain private and accessible to the residents only.

User feedback

The houses are thermally comfortable and users do not have to use fans till mid-April when the mean ambient temperature is around 25 °C, the maximum day temperature being about 35 °C.

At a glance

Project details

Project description This cooperative housing project consists of 180 houses on individual plots. The project attempts at creating houses suited to the needs of the individual owner.

Location Gurgaon

Owner National Media Centre Co-operative Housing Society Ltd

Climate Composite

Design team Vinod Gupta, L P Singh, Anita Narula, Harvinder Singh, Alka Arora

Consultants

Structural Kirti Consultants

Plumbing R K Gupta & Associates

Built-up area 35 000 m²

Completion 1996

Electrical Electrical Consulting Engineers

Landscape Parivartan

Project cost Rs 170 million

Design features

- Choice in housing where people's requirements are matched to their house plans
- Houses designed to be thermally comfortable and energy-efficient
- Roofs insulated with earthen pots laid in mud phuska
- Each house equipped with a solar chimney where a single evaporative cooler can be fitted by minimal ducting, the entire house can be cooled
- All houses provided with plumbing for fitting solar water heaters
- Garbage to be sorted within the house into biodegradable and non-biodegradable matter and the organic matter is to be composted by vermiculture and the inorganic matter sold
- The wiring in each house has been done such that an inverter can be easily installed
- Installed renewable energy systems and waste management techniques include rainwater harvesting, solid waste recycling, and solar water heating

Temperature check for Gurgaon

(see Appendix IV)

Cold	Cool	Comfortable	Warm	Hot	
	●				January
	●				February
		●			March
			●		April
				●	May
				●	June
			●		July
			●		August
			●		September
		●			October
		●			November
	●				December

Add the checks

	3	3	4	2

Multiply by 8.33% for % of year

Heating	25
Comfortable	25
Cooling	50

American Institute of Indian Studies, Gurgaon

Editor's remarks

With an impression of a small single-storeyed structure from outside, the building opens up into a series of functional courtyards, gardens, pavilions, which have been traditionally used as climate modifiers in Indian architecture. Earth berming, proper orientation, and fenestration design are a few passive solar techniques used to reduce air-conditioning loads. Innovative space-conditioning techniques and use of renewable energy systems further reduce the pressure on grid electricity.

Architect Vinod Gupta

Ethos of traditional Indian architecture reflected in the design of a modern building to achieve energy efficiency and environment-friendliness

The AIIS (American Institute of Indian Studies) is a consortium of American universities that provides scholars with facilities for research in Indian art, architecture, and music. The new building houses the administrative offices of the Institute and the research facilities, archives, and libraries of the Centre for Art and Archeology and the Centre for Ethnomusicology. From the outset, the building was planned to be energy-conserving and environment-friendly.

▲ With an exposed brick finish, the building has staggered walls and the walls provide mutual shading. The building has an appearance of a small single-storeyed structure from the outside and holds the visitors by surprise as one travels across the interior spaces

The concept—fusion of traditional architecture with modernism

One of the differences between traditional Indian architecture and European-style buildings is in the diversity of functional spaces that an Indian building provides in the form of gardens, courts, verandahs, pavilions, and passages in addition to the normal enclosed rooms. In contemporary Indian architecture, these elements are often used merely for their formal effect. Non-functional *chhatris* in air-conditioned buildings top the list of misused architectural forms. Open and partially covered or enclosed spaces are of special significance in traditional Indian architecture. Gardens, courtyards, pavilions, and verandahs allow people to be comfortable at those times when they don't feel the same in the fully enclosed rooms. The AIIS building follows this tradition of using spaces with different thermal characteristics.

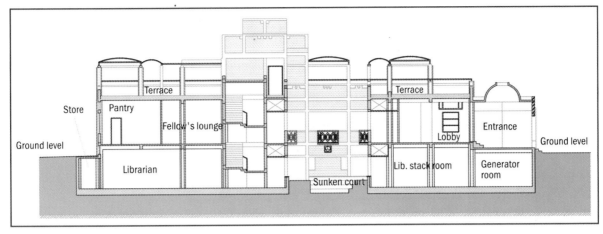

▲ Section of building showing partial sinking of the building into the ground to take advantage of the thermal storage capacity of the earth

Passive solar techniques

Courtyard planning and terrace garden

As the AIIS was set up to conserve Indian art, architecture, and music, the building design refers to traditional Indian way of building without trying to be monumental. From outside, the building appears to be a small single-storeyed structure. As one moves in from the lobby, the building opens up to reveal two sunken courtyards with vegetation and water. The sunken courtyards provide an element of surprise to the visitor and bring adequate daylight into all areas of the building including the basement and are designed for holding small meetings in summer. The architect has attempted to restore to the courtyard its original function of being an internal climate modifier rather than being simply a light well.

The terrace garden with its modern pavilions is suitable for larger gatherings, which usually take place in winter. The breezy central verandah is ideally suited for the monsoon period. The first court is ornamental with a water pool and fountain while the second one has plants and sitting areas. The diagonal placement of two courtyards, with the second one being open-ended, has immensely increased the air circulation within the building.

Earth contact

Studies in various parts of the world have shown that the feeling of comfort or discomfort depends on people's expectations much as it depends upon the actual indoor temperature. So it was desirable to have the feel of a naturally cooled building with evaporative cooling installed as an extra rather than the feel of a fully enclosed modern building in which the air-conditioning is yet to be installed. Since it is difficult to cool a multi-storeyed building by passive solar techniques, in the new design, the building was limited to basement and ground floors. The workspaces are arranged around two sunken courtyards with water and plantation. The passages around the courtyards form a thermal transition zone between the outdoor temperature and the controlled indoor temperature, modulating people's

▲ Internal courtyard as a climate modifier and daylight well for lighting up the spaces

expectation of thermal comfort as they move from one space to another. The building form emulates the peaceful internal environment of a traditional courtyard building while maintaining the modern standards for natural lighting and ventilation.

The archives have to be maintained at a temperature of 18 degrees and need to be air-conditioned anyway. Some of the associated technical areas also require air-conditioning. To reduce the cooling load, these special areas and the libraries are located in the basement. The work areas located at the ground floor are protected by specially insulated (expanded polyethylene) external walls and cooled by a terrace garden.

Orientation

The building façade is staggered at 45 degrees to the site boundary which allows north–south orientation for the windows thus reducing heat gain through windows.

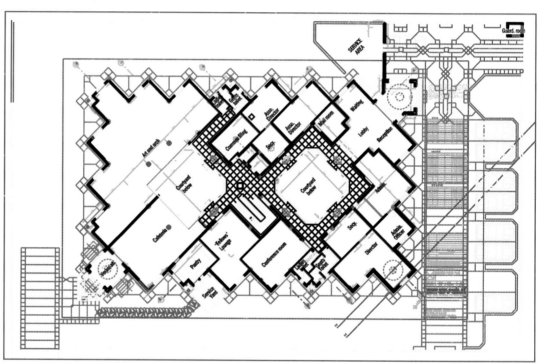

Building plan showing walls angled at 45 degrees to the site boundary, which allows north–south orientation for the walls thus reducing heat gain through windows

Fenestration and shading

The windows have adequate sun-shading to reduce direct solar gains. On the east–west faces, small slit-like openings are filled with glass brick to let in light without heat and to improve the distribution of light within the room. Light shelves, in the form of white ceramic tiled surfaces, on top of the sunshades and below the ventilator, increase light penetration into the interiors by reflecting the external light on to the ceiling. Long Corbusian louvred slits on the internal wall of the director's room increase air circulation and light. All these devices ensure that no artificial light is needed during the day.

Water heating

Solar water heating for 250 litres per day has been provided for in the cafeteria and the staff quarters.

Innovative space cooling concept

Gurgaon has a composite climate like that of Delhi and evaporative air-cooling is effective only for about two months in a year. In this climate, passive solar architecture cannot provide the kind of thermal comfort that air-conditioning does. When the initial design of the building was costed, the estimated cost of Rs 30 million rupees exceeded the financial resources of the Institute by a factor of two. By reducing the built areas, it was possible to reduce the cost to Rs 25 million rupees and no further. To build the area required within the available money required drastic changes in design and this is where energy-saving techniques were found to be useful.

It was found that air-conditioning was a major cost element not only as direct capital cost of air-conditioning plant, but also as the cost of electrical sub-station equipment, generators for standby power, and the built-up space required for the air-conditioning plant, and electrical equipment. It was calculated that if air-conditioning was replaced by evaporative air-cooling, the total connected power load of the building could be reduced to about 60 kW. At this level of power demand, the electrical sub-station would not be required and a small generator would suffice for the standby power requirements. The cost of the entire project could then be reduced to less than Rs 20 million rupees. However, the change to air-cooling, which adds humidity, is not generally acceptable for archives unless some special measures are adopted. Therefore, a combination of both passive solar architecture and a special two-stage evaporative cooling was adopted.

The first stage provides normal direct evaporative cooling in which air is cooled by addition of moisture. In the second stage, air is cooled indirectly by passing it over a heat exchanger carrying cooled water, without the addition of humidity. By controlling the operation of these two stages, it is possible to achieve cooling with some degree of humidity control. Ducts carrying cooled air run over the passages around the courtyards and blow cooled air into the work areas.

Traditional flavour to interiors

The traditional flavour of the building continues in the interior design also. The teakwood furniture looks traditional but is designed for use with computers. Traditional crafts are visible everywhere. Terracotta relief done by a master craftsman have been used as decorative elements in the courts. Textiles from various parts of India have been used as decorations. The decorative motifs used in the flooring, the railings, and in some of the furniture are all derived from traditional Indian designs. Even the large sand-cast relief of the sun god, created by P Daroz, a modern artist, has the appearance of traditional crafts. The building is finished with exposed brick of terra cotta colour interspersed with beige firebricks. The domes over the entrances are clad with green and black glass mosaic.

Post construction scenario and performance

The archives, libraries, conference areas, and some of the office areas needed air-conditioning for maintaining temperature and humidity round the year. However, the loads of these areas have been substantially reduced by carefully locating these areas in the basement or by providing sufficient insulation.

The building uses direct/indirect evaporative cooling system for dry summer (excepting in the archives where air-conditioning is used round the year) and switches over to air-conditioning partially (about 60% of the area has provision for air-conditioning) during humid summers. The building is primarily a day-use building with a five-day working week, with the archival spaces air-conditioned throughout day and night all round the year. The total installed capacity of the building is now 139 kVa. The building consumes about 12 500 units of electricity per month during peak summers and 7500 units of electricity per month during peak winters. Well lit round the year, the building rarely uses artificial lights during daytime and the unconditioned spaces are also comfortable round the year, according to the occupants of the building.

At a glance

Project details

The AIIS is a consortium of American universities that provides facilities for research in Indian art, architecture, and music

Building type Institutional
Location Gurgaon
Climate Composite
Architects Vinod Gupta, L P Singh, Amrita Sharma, Koheli Banerjee of Space Design Associates, New Delhi
Floor area 1500 m²
Year of completion 1998
Cost Rs 20 million

Consultants
Structure Y N Diwan
Mechanical Gupta Consultants
Elecrical Engineering Consulting Services
Plumbing R K Gupta
Landscape Kavita Ahuja
Main contractor Parshotamsons Construction
Furniture Sharma Construction

Design features

- Building façade staggered at 45 degrees to the site boundary allows north–south orientation for the windows thus reducing heat gain through windows
- Adequate shading for north–south windows to reduce direct solar gains. Small slit-like openings on the east-west faces, filled with glass brick to let in light without heat and to improve the distribution of light within the room
- Light shelves, in the form of white ceramic tiled surfaces, on top of the sunshades and below the ventilator to increase light penetration into the interiors by reflecting the external light on to the ceiling
- Courtyards as microclimate modifier and for daylighting
- Roof garden over the work areas to reduce space-conditioning loads
- Expanded polyethylene as wall insulation for external walls
- Earth shelter moderates internal temperatures
- Two-stage evaporative air-cooling for summer with minimum air-conditioning for the critical areas requiring humidity control
- Solar water heating in the cafeteria and the staff quarters

Temperature check for Gurgaon

(see Appendix IV)

Cold	Cool	Comfortable	Warm	Hot	
	●				January
	●				February
		●			March
			●		April
				●	May
				●	June
			●		July
			●		August
			●		September
		●			October
		●			November
●					December

Add the checks

	3	3	4	2

Multiply by 8.33% for % of year

Heating	25
Comfortable	25
Cooling	50

Climatic zone: hot and dry

Indian Institute of Health Management Research, Jaipur

Sangath – an architect's studio, Ahmedabad

Torrent Research Centre, Ahmedabad

Residence for Mahendra Patel, Ahmedabad

Solar passive hostel, Jodhpur

Indian Institute of Health Management Research, Jaipur

Architect Ashok B Lall

Expressions to seek modernity in an evolving traditional culture and to express dignity and richness through deployment of very simple, energy-efficient and economical means

> **Editor's remarks**
>
> An architectural expression inspired by the traditional arts and crafts of Jaipur, the building maximizes use of local resources to effect economies in costs on the one hand, and reduces transportation energy on the other. Use of passive architectural principles suitable for hot, dry climate of Rajasthan further adds to energy efficiency.

The building of the IIHMR (Indian Institute of Health Management Research) in Jaipur has been inspired by the traditional arts and crafts of the city. The architectural vocabulary is derived from the use of a variety of stones for masonry, paving, coping, flooring, and pergolas in conjunction with slender precast concrete elements.

Located near the airport at Sanganer on the outer reaches of Jaipur, this Institute, the only one of its kind in India, is devoted to research and training in the management of health systems. The IIHMR project was the winning design entry of a closed competition held in 1988.

▲ Entrance forecourt with fountain. The fountain acts as a modifier of microclimate and moderates ambient conditions in the hot dry climate of Jaipur

Site planning

The natural topography of the land provided the rationale for the site planning of the institutional and residential components. The seasonal drainage channel was visualized as a leisure valley. Courtyards and terraces were to ride the slopes and retain continuity with the land beyond the site. Low-lying and less water-starved land was to become parkland and the residential and institutional buildings were placed across it. The leisure valley was thought of as the primary focus for informal interaction for the entire community with

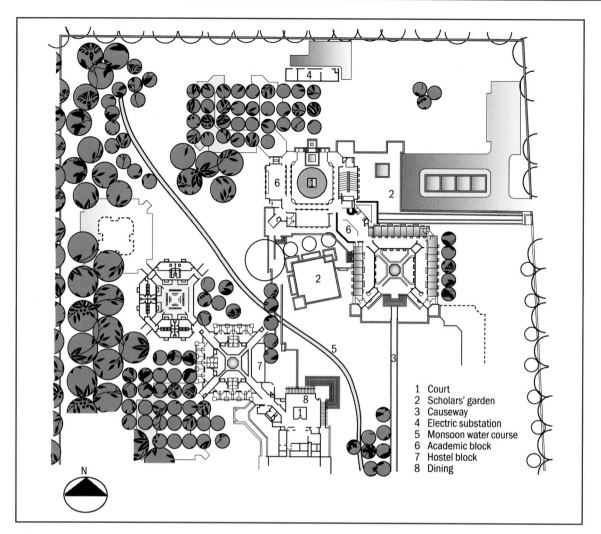

▲ A series of interlinked courtyards in the IIHMR. Compact planning with a series of interlinked courts provides a well-understood language to convey the social and functional structure within the campus community

stepped ghats, shade, and flowers. A bridge at the centre of the site crosses the valley between the two sets of buildings, while a causeway at the eastern edge of the valley acts as a dam to impound rainwater. This helps in recharging the water table apart from adding to the environmental character of the campus.

Courtyard planning of individual clusters

A second characteristic element was the compact planning of the buildings around linked courts. The nature of these courts resulted from an analysis of the primary activities of the Institute, which were seen to consist of study and interaction. It was by structuring these two activities into a programme of work that the Institute's further objectives of research, training, and dissemination are served.

In the competition scheme, the courts were planned with their own individual characters. A study court forming a community of scholars links to a college court expressing interaction (lectures, workshops, etc.) within the Institute by encouraging contact between various groups. This, in turn, was to be linked to a public court formed out of the spaces where the Institute communicates with the world at large through exhibitions, films, conferences, and the like.

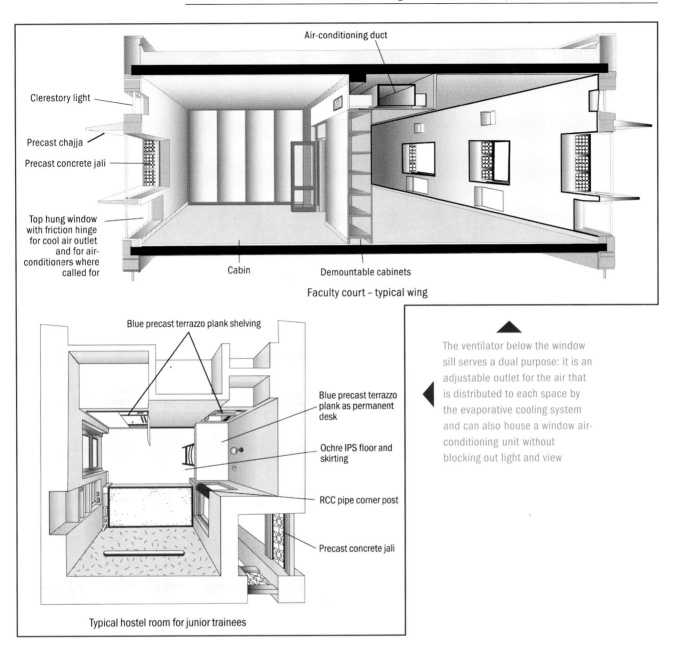

Faculty court – typical wing

The ventilator below the window sill serves a dual purpose: it is an adjustable outlet for the air that is distributed to each space by the evaporative cooling system and can also house a window air-conditioning unit without blocking out light and view

Typical hostel room for junior trainees

As it happened, the requirements were greatly curtailed. The courts were reworked as a 'faculty court', which brings together research and administrative functions, and an 'academic court' around which the training functions are organized. Further reductions in area led to only three sides of the court being taken up as Phase I.

Each residential courtyard expresses a community. The two hostel courts planned so far provide accommodation for long-term trainees around one, and for visiting faculty and senior training participants around the other. The trainees' hostel accommodation is in the form of flats. In each flat, three small study-cum-bedrooms share a common lounge and toilet facilities. The organization of the faculty court allows a modular variation in the size and number of the rooms with removable partitions. Similarly, the classroom spaces in the academic court are designed on a grid, which permits the formation of a variety of classroom sizes.

Window design

The pattern of fenestration is also coordinated with the planning grid. The windows are designed to reduce glare, and yet ensure adequate natural light for all workspaces. The small high-level glazing panel supplements window light by throwing more daylight at the back of the room to give a fairly even illumination level across the depth of the rooms. The ventilator below the window sill serves a dual purpose: it is an adjustable outlet for the air that is distributed to each space by the evaporative cooling system and can also house a window air-conditioning unit without blocking out the light and view.

Space cooling

The cooling plant and service cores are designed to ensure a noise- and draught-free air cooling system. With the exception of a few air-conditioned rooms, all workspaces are served by a built-in evaporative cooling system. Jaipur being a hot, dry area, this system provides a high level of comfort at very little cost.

▲ Picture showing design of window

Materials

The essence of the expressive qualities of the Institute is found in the stone used here. And it is here that we can begin to unravel some of the interaction between design thought and its implications when applied in practice. The competition design report sets out some design concerns. As Jaipur possesses one of the finest traditions of the craft of building, using locally available skills and materials, while being inherently economical, would also serve the important objective of preserving and strengthening a fine building tradition. The structure was, therefore, largely based on load-bearing masonry construction. There was a conscious effort to select construction methods and techniques that exploit the characteristics of stone construction. Its expressive qualities of colour, texture, and exposed finish were harnessed to give character to the building. It is also a low-maintenance finish.

The architects made a rapid survey of the skills available to them. They identified stone quarries, local craft-based workshops such as jali-makers and other construction-based resources such as a precast concrete workshop of a good standard. Stones available in the quarries were identified. Using different kinds of stone as tradition (on the basis of their individual workability) was considered. Clearly, one can observe and appreciate how a larger framework of attitudes towards economy of means, appropriateness, and the preservation of an ancient building tradition guided the architect, whose experiences in applying these notions in this project are of interest.

First, a sample stone was built on site before tenders were awarded. Contractors quoting for the work would, therefore, see the standard and quality required.

Second, the precast concrete jali work and terrazzo shelves were to be awarded to the workshop selected by the architect as a nominated subcontractor on rates agreed upon. Such steps ensured that difficulties faced by the contractors could be overcome to meet design intentions.

To take just one example, the stone chosen, an attractive pink quartzite selected at the quarry for its colour and expressive quality, was normally handled by less-skilled masons. This was because the stone was considered too 'coarse' to attract the more-skilled artisans. The contractor's masons had to be trained to build to the standards set. The polygonal stone masonry was disciplined by course-lines at designed intervals. Also of note is a controlled irregularity in texture and colour.

Energy conservation

Energy is sought to be conserved on many fronts. Load-bearing construction reduces the use of steel and cement in the building structure. The massive construction provides thermal dampening for both summer and winter. All workplaces receive adequate natural light during clear days. Each building wing has a built-in evaporative cooling system, which provides thermal comfort at low energy costs. Courtyard with fountains modifies the microclimate.

At a glance

Project details

Project Indian Institute of Health Management Research, Jaipur
Building type Institutional
Climate Hot and dry
Area 5500 m²
Ccst Rs 20 million
Year of completion 1991 (Phase 1)
Client Society for Indian Institute of Health Management Research, Jaipur
Design team Ashok B Lall, Rakesh Dayal, Aditya Advani, Sunita Sharma, Sujata Kacker, Neha Kulkarni, Vineeta Gothaskar
Consultants Engineering Consultants India (structural), Spectral Service Consultants (electrical and HVAC) Deolalikar Consultants (public health), M R Mehendale (quantity surveyor), Deepak Hiranandani (landscape)
Contractors Gurbaksh Singh (civil), Anita Electricals (electrical), Suvidha Engineers (HVAC)

Design features

- Interlinked courtyards help in increase of heat loss by ventilation, which also form functional spaces for the facility
- Landscaped courtyards used as a modifier of microclimate
- Most of the spaces are air-cooled using centralized evaporative cooling system, which provides a high level of comfort in hot dry climate at very little cost
- Windows are designed to reduce glare and ensure adequate natural light for the entire room
- Use of local materials and skills reduces costs and also reduces transportation costs
- Interconnecting walkways between various blocks, shaded using natural vegetation cover provides a cooling effect
- Innovative window shading using concrete jali as side fins, provides shading and yet does not impede air movement

Temperature check for Jaipur

(see Appendix IV)

Cold	Cool	Comfortable	Warm	Hot	
	●				January
		●			February
		●			March
			●		April
				●	May
				●	June
			●		July
			●		August
			●		September
			●		October
		●			November
		●			December

Add the checks

	1	4	5	2

Multiply by 8.33% for % of year

Heating	8
Comfortable	33
Cooling	58

Sangath – an architect's studio, Ahmedabad

Architect Balakrishna Doshi

A bioclimatic response to hot, arid climate

Editor's remarks

Sangath – spatial, constructional and landscape response to combat hot and dry climate of Ahmedabad. Various passive solar architectural techniques have been adopted to negate the impact of harsh sun.

The relative organization of form elements, layering of spaces, controlled interiors and transitions to outside, the interruption of the skyline through varying outlines that break the sun into shadow, and open the roof into the night sky are themes that respond well to the hot and arid climates. At Sangath, the architect's design studio, these responses are at their best. The architectural studio comprising reception areas, design studio, office spaces, workshop, library, conference room, and other ancillary spaces has been designed to naturally manage the forces of nature. There are spatial, constructional, and landscape responses to combat the vagaries of nature in the hot dry climate. In Ahmedabad, the summer temperature reaches up to 45 °C and the heat is very intense. It is the heat rather than breeze that becomes critical. Natural comfort conditions can be achieved by protecting spaces from the heat and glare of the sun.

▲ A view of China mosaic clad vaulted roofs of Sangath which reflect heat, reducing heat gains to internal spaces

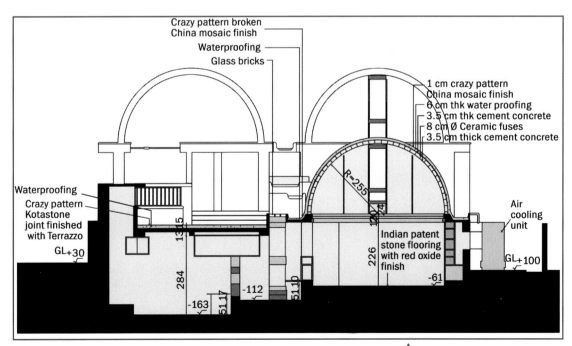

A section through Sangath showing subterranean spaces with vaulted roof creating ▲ efficient surface/volume ratio thereby optimizing material quantities

Design responses to hot dry climate

Subterranean spaces

The building is largely buried under the ground to use earth masses for natural insulation.

Storage walls

External walls of the building are nearly a metre deep but have been hollowed out as alcoves to provide storage that becomes an insulative wall with efficiency of space (for storage functions).

Laying of locally-made clay fuses over concrete slab during the sandwich vault construction stage. The clay fuses entrap air thus forming a non-conductive layer. This vault has then been topped using broken China mosaic tiles, which reflect heat and retard heat transmission.

Vaulted roof form

The roof form creates efficient surface/volume ratio optimizing material quantities. The higher space volume thus created provides for hot air pockets due to convective currents that keep lower volumes relatively cool. The ventilating window at upper volume releases the accumulated hot air through pressure differences.

Sandwiched construction of vault

The vaulted roof is of locally-made clay fuses over the concrete slab, which provides a non-conductive layer. The top finish of China mosaic glazed tiles further enhances insulation by being white and glossy to reflect sun and being of clay to retard heat transmission.

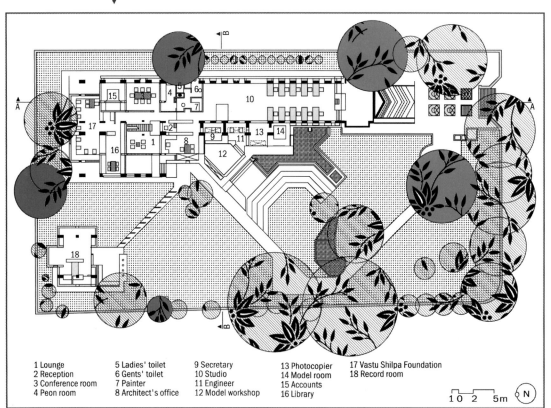

The ground-floor plan

1 Lounge
2 Reception
3 Conference room
4 Peon room
5 Ladies' toilet
6 Gents' toilet
7 Painter
8 Architect's office
9 Secretary
10 Studio
11 Engineer
12 Model workshop
13 Photocopier
14 Model room
15 Accounts
16 Library
17 Vastu Shilpa Foundation
18 Record room

Water is a major modifier of the microclimate. Rainwater and overflow of roof tank are harnessed through roof channels and reused. Water also moderates the harsh ambient conditions near the building

Indirect/diffused light

Sun light brings heat and haze with it. To maximize daylight (intensity of illumination) and to diffuse heat and glare, the light is received in indirect manner by diffusing it. There are three ways by which natural light is drawn within.

- By upper-level large openings towards north direction, which is cool, and consistent light is reflected off the clouds
- Skylights, which are projected masses from the roof, reflect the light on the white inner wall surface, which further radiates light in the room.
- Innermost spaces are lit up through small cutouts in the roof slab, which are then filled with hollow glass blocks that take away the glare and transmit diffused light

Microclimate through vegetation

Lawn and vegetation cover all around create favourable microclimate by absorbing solar radiation and providing cooler passage of air through humidity.

Water channels

Rainwater and overflow of pumped water from the roof tank are harnessed through roof channels that run through a series of cascading tanks and water channels to finally culminate in a pond from where it is recycled back or used for irrigating vegetation. Water cascades also provide interesting visual experiences.

Exposed natural finishes

Concrete of slab and wall surfaces are kept bare unplastered as final visual finishes, which provide its own natural look and save on finishing material quantity.

Use of secondary waste material

Paving material is a stone chip waste while roof surface is glazed tiles waste, which are available as waste material from factories at no cost. These have been creatively hand-crafted and integrated in design by fully using waste material. The application is also skill-oriented and involves as well as promotes craftsman and our traditional heritage.

Performance of the building

The above measures have ensured excellent climate control within in terms of keeping inside cool and increasing time-lag for heat transfer. There is a difference of about 8 °C between interior and exterior of the roof skin temperatures.

The time-lag for heat transfer is nearly six hours. The natural elements are harmoniously blended with the built environment and water recycling and waste material reuse have ensured cost economy as well as environmental consciousness.

The exposed surfaces have saved nearly 10% of the project cost usually spent on finishes. Water recycling has been most rewarding economically to keep lawn areas possible. Natural daylight ensures minimum electrical consumption for artificial light and all insulative measures have resulted in nearly 30% to 50% cost reduction in cooling energy.

At a glance

Project details
Building/project name Sangath –an Architect's Studio
Location Ahmedabad
Building type Institutional
Architect Balkrishna Doshi
Climate Hot and dry
Year of start/completion 1979–1981
Client/owner Balakrishna Trust
Site area 2346 m²
Covered area 585 m²
Cost of the project Rs 600 000 (1981)

Design features
- Underground construction
- Thermal storage walls
- Vaulted roof form to create efficient surface/volume ratio. The vault induces convective air movement thereby cooling internal spaces
- Vaulted roof of sandwiched construction with an insulating layer of locally made clay fuses sandwiched between two concrete slabs
- Use of broken China mosaic glazed tiles from local factory as top finish for the vault to reflect heat
- Daylighting by north-glazing, skylights, and roof cutouts
- Microclimate modified by vegetation and water bodies
- Rainwater and roof tank overflow water harnessed for recycling and reuse

Temperature check for Ahmedabad
(See Appendix IV)

Cold	Cool	Comfortable	Warm	Hot	
		●			January
		●			February
			●		March
			●		April
				●	May
				●	June
			●		July
			●		August
			●		September
			●		October
		●			November
		●			December

Add the checks

Cold	Cool	Comfortable	Warm	Hot
		4	6	2

Multiply by 8.33% for % of year

Comfortable	33
Cooling	67

Torrent Research Centre, Ahmedabad

Architects Nimish Patel and Parul Zaveri

An attempt to find technological solutions to the power problems that the country faces

Editor's remarks

In a world where developed societies with their sustained focus on increasing importance of individuals have caused excessive consumption of electrical energy for cooling internal spaces, the Torrent Research Centre is a welcome experiment. It demonstrates that innovative technological solutions drastically cut down space-conditioning and artificial lighting loads in a building without compromising either on required levels of thermal and visual comfort or on recommended indoor air quality. The architectural language expresses the inherent character of building functions and results in an imposing and attractive appearance, which according to the architect has a 'quality of timelessness'.

In this project, architects at Abhikram have attempted to extend the use of natural light, ventilation, and cooling for human comfort to large-scale projects—an approach that was accepted by the clients. The clients were convinced that the experiment, if successful, would be a major step forward in the direction of energy conservation. When the proposal was initially mooted, the clients were skeptical, but it became increasingly interesting and even challenging.

The TRC (Torrent Research Centre) is a complex of research laboratories with supporting ancillary facilities and infrastructure located on the outskirts of Ahmedabad. The complex comprises almost all the disciplines of pharmaceutical research with their range of areas—the cleanest requiring Class 10000 atmosphere, to the dirtiest emitting obnoxious gases. The 30-acre relatively flat site is located in the vicinity of major institutional buildings. The brief was well defined with the need to optimize overall project costs so as to achieve a balance between all the needs, with the worth of the money spent fluctuating within five per cent.

The architects attempted to use innovative approaches that strike a satisfactory balance between the varied requirements without compromise. The design, therefore, maximizes the use of locally available natural materials, minimizes the use of artificial light, and attempts to minimize the use of conventional air-conditioning with the introduction of the PDEC (passive downdraft evaporative cooling) system. Further, the aesthetic language represents the inherent character of the building.

▶ Exhaust shafts, a building language

◀ Night-time view of the building

The TRC is the result of the efforts of Abhikram which started with the involvement of Brian Ford & Associates for the design of the typical laboratory building. After going through several solutions, including the ground cooling, the present section of the typical laboratory block was jointly evolved by them. Consequently, the administrative block and the laboratory core block were designed by Abhikram but vetted for their sizes, heights, and volumes by the Solar Agni International, Pondicherry.

Passive downdraft evaporative cooling system

The design was aimed at integrating spaces requiring highly controlled conditions with those requiring less-controlled conditions while minimizing the presence of dust in the internal environment. Passive cooling is attempted through a system of designated inlets and outlet shafts.

The shafts as a consequence of their locations, sizes, heights, and their complex but simulated and in-depth researched configuration generated the required movement of air in different spaces without using any mechanical or electrical energy. Thus the buildings have a climatically-sealed environment with only designated inlets and outlets, with air cooled at the point of entry using a fine spray of water.

The buildings have been under observation since they were first occupied and will continue till 2001. Observations so far indicate that in the three principal seasons (summer, monsoon, and winter) since occupation, human comfort conditions have not been compromised. Further observations include the following.

- In the summer of 1997 when the outside temperature registered 43–44 °C, the inside temperature remained at 29–30 °C—a 13–14 °C drop.
- Six to nine air changes per hour were noted on different floors, including in a chemistry laboratory.
- Temperature fluctuations inside the building have not exceeded beyond 4 °C over any 24-hour period. At the same time, the temperature fluctuations outside were as much as 14–17 °C.
- Very rarely did the inside temperature fluctuations cause any level of discomfort to occupants.

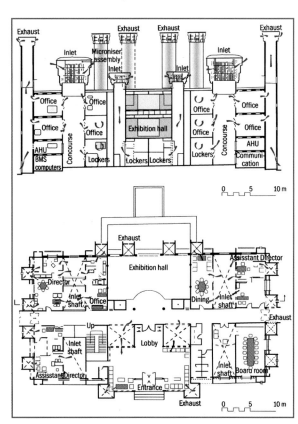

▲ Plan and section of administration block

Materials, techniques, and methods

The RCC-framed (reinforced cement concrete-framed) structure, with brick in-filled walls, has glossy enamel paint on cement/vermiculite plaster on the internal surface, and textured cement plaster on the exterior. Flooring and other finishes have been selected on the basis of functional requirement. To minimize the collection of dust and cobwebs at junctions and corners, internal plastered surfaces are either curved on the inside or the outside corner edges. Greater dust and cobweb control has been achieved by eliminating down-turned beams from clear spaces, using hollow concrete blocks on flat form work forming voids in which the RCC coffered slab is cast. Vermiculite is extensively used for insulation in roof and cavity walls, along with cement brick-bat-based water-proofing technique.

This will help achieve the required R-values and reduce the chemical disharmony of adjacent materials. Half-rounded ceramic pipes on the outer face of the inlet and exhaust shafts of the PDEC system create local turbulence, thus reducing the entry of large dust particles. Other dust reduction mechanisms, provided for but not executed, include universal louvres and insect screens. Motor-operated shut-off louvres were provided for dust storm conditions, but were removed later.

▲ Daylit internal corridor of the Torrent Research Centre

Economic projections

The economic viability of the project is demonstrated by the following indicators projected for the total project on the basis of results from buildings under observation.

- The additional cost of civil works including insulation is about 12%–13% of conventional building.
- Saving on air-conditioning plant and equipment is about 200 TR (tonnes of refrigeration)—about 65% of the additional cost mentioned above.

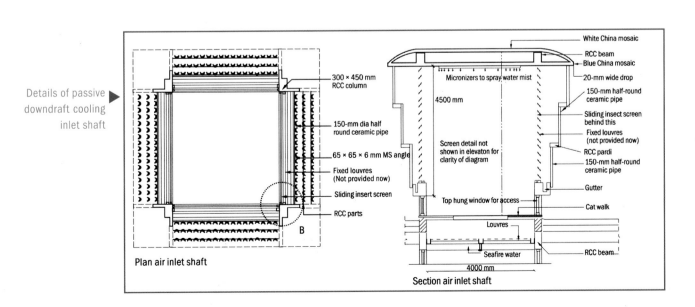

Details of passive downdraft cooling inlet shaft

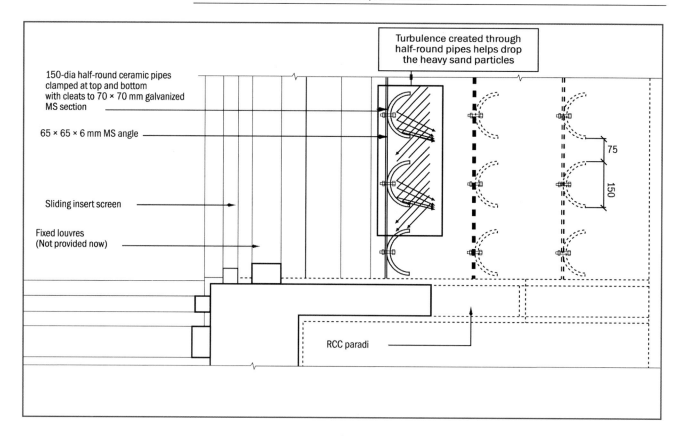

- Cumulative capital cost of civil works and air-conditioning plant is approximately Rs 5 million more than that of conventionally designed buildings.
- Annual savings in electricity consumption including savings on account of non-use of artificial lighting during the day will be approximately Rs 6 million.
- Payback period of the total capital cost from the saving of electrical consumption alone will work out to about one year. The payback period for the cost of the entire complex from the saving on electrical consumption as well as plant replacement costs will work out to around 13 years.

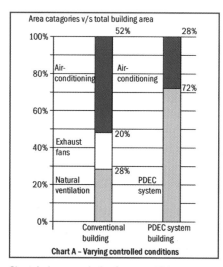
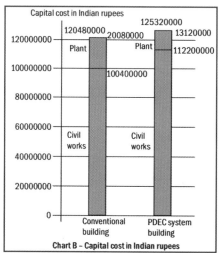
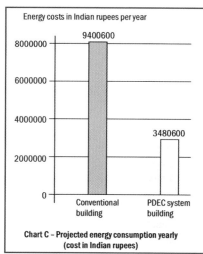

Chart A shows analysis of areas, which are shifted to PDEC system from conditioning system without compromisng the human comfort level

Chart B shows net increase of Rs 4 840 000 in the capital costs if civil works and AC plants by adopting PDEC system

Chart C shows the projected net energy savings of Rs 5 920 000 per year by using PDEC system

Note Eco-friendly approach to design has saved energy cost by Rs 5 920 000 without sacrificing the human comfort; in our energy-starved nation, making this energy available to other areas of the economy has a far greater and immediate, direct and indirect, impact on the environment

Scientific observations of the building's behaviour

The TRC complex has been under observation, using elaborate procedures of recording the data. Some of the data have been reproduced here to explain the depth of the attempts to understand the behaviour of the building under observation.

Chart A shows analysis of areas that are shifted to the PDEC system from the air-conditioning system without compromising the human comfort level; Chart B shows net increase of Rs 4.85 million in the capital costs of civil works and air-conditioning plants by adopting the PDEC system; Chart C shows the projected net energy savings of around Rs 6 million per year by using the PDEC system.

At a glance

Project details

Project description The Torrent Research Centre is a complex of research laboratories with supporting facilities and infrastructure located on the outskirts of Ahmedabad.

Architects Nimish Patel and Parul Zaveri, Abhikram, Ahmedabad

Energy consultants Brian Ford, Brian Ford and Associates, London, UK (for the typical laboratory block in all aspects); C L Gupta, Solar Agni International, Pondicherry (for the rest of the blocks, vetting Abhikram designs)

Project period 1994–1999

Climate Hot and dry

Client/Owner Torrent Pharmaceuticals Ltd

Size Built-up area of approximately 19 700 m²

Builder/Contractor Laxmanbhai Constructions (India) Pvt. Ltd; M B Brothers Ltd; Shetusha Engineers and Contractors Pvt. Ltd; Materials Corner; J K Builders

Materials, techniques, and methods

- Design maximizes the use of locally available natural materials and avoids the use of synthetic materials.
- RCC-framed structure with brick in-filled walls, with glossy enamel paint on cement/vermiculite plaster on the internal surface.
- All internal plastered surfaces are either curved on the inside or curved on the outside corner edges to minimize dust collection and cobwebs at the junctions and corners.
- For greater dust and cobweb control, the downturned beams are eliminated from the clear spaces, through a system of constructing the RCC slab, using hollow concrete blocks on the flat form work, forming voids in which the RCC coffered slab is cast.
- Vermiculite, a natural mineral, is extensively used for the insulation in roof and cavity walls to achieve the required R-values, along with cement–brickbat-based waterproofing
- PDEC system has been designed and adopted for space conditioning of the building.
- Daylight integration has been made for reducing energy usage.
- Innovative use of half-round ceramic pipes, on the outer face of the inlet and exhaust shafts of the PDEC system, to reduce the entry of larger dust particles by creating local turbulence.

User statement

When the proposal was initially mooted we were sceptical but decided to listen to what our architects had to say. As the story unfolded and drawings made, it became more interesting and even exciting to think of a system that would use little energy and yet provide comfort for almost 9–10 months in a year. We had nothing to lose; the additional cost of the construction versus the saving on O&M of plant and equipment for almost 200 TR (tonnes of refrigeration) of cooling was attractive.

Summer of 1997 was the acid test of one of the buildings and the results were very gratifying. We had held up placement of fans in the labs to know the reactions. There were no complaints. No one felt the need for one. The labs were comfortable to work in without fans. They were not stuffy or smelly as most chemistry labs are, even when air-conditioned. This was the added advantage.

Monsoon, as expected, was not so comfortable, the labs were muggy and we had to install fans to provide comfort for these 2–3 months. There was some reverse flow of air from the exhausts to the inlets on days that were windy.

Winter again has been comfortable as I am sure summer would be too. The first experiment has worked well and hopefully this augurs well for the Torrent Research Centre.

Director, Torrent Research Centre
February 1998

Temperature check for Ahmedabad

(See Appendix IV)

Cold	Cool	Comfortable	Warm	Hot	
		●			January
		●			February
			●		March
			●		April
				●	May
				●	June
			●		July
			●		August
			●		September
			●		October
		●			November
		●			December

Add the checks

		4	6	2

Multiply by 8.33% for % of year

Comfortable	33
Cooling	67

Residence for Mahendra Patel, Ahmedabad

Architect Pravin Patel

Editor's remarks

Compact planning, use of appropriate materials of construction, wall and roof insulation, and double glazing have drastically reduced the space-conditioning loads in this residential building. Part of the building load is met by judicious use of renewable energy systems. Use of building automation further enhances energy savings in the building.

Here is one house that depends on solar energy to a great extent. Minimally relying on the grid for power, the architect has also integrated the house with an automation system for the purpose of saving energy

▲ The east façade, which happens to be the front, is shaded by large overhangs. The overhangs also act as service ducts and thus serve two purposes. The white texture of the house reflects incident radiation

Building design

The residence of Mahendra Patel (client), located in the hot and dry climate of Ahmedabad, was required to be fully air-conditioned as per the requirement of the client. However, the client was extremely concerned about the energy usage in the building and wanted to achieve an energy-efficient design along with optimized use of solar energy for meeting a part of its electrical requirements and water heating needs. The architect has tried to minimize the air-conditioning load by applying passive solar design interventions. Judicious use of solar energy systems has been made to meet a part of the energy demand in the building load.

Ahmedabad is located in the 'hot dry' zone where the summer temperature goes up to 45 °C. Natural comfort conditions can be achieved by protecting spaces from the heat and glare of the sun. The site of about 755 m² is oriented east–west with the shorter side facing the road.

The house is designed in such a way that all the private spaces on the ground and first floor are connected to the entrance hall through individual doors. This helps in reducing passage area and loss of air-conditioning. The double height of the entrance hall gives a grand look. Though the width of the plot is less, the rooms do not look small because of bay windows and angled walls. The verandahs play an important role in the Ahmedabad climate. Hence, the house has been provided with two big verandahs.

Design, materials, and construction techniques

Building envelope

The building has adequately designed windows with overhangs on south wall. Thermally massive stone floors and thick masonry walls moderate diurnal variations. East and west walls are protected by a series of extended terraces developed as building elements to shade walls. Double glass shutters reduce heat gain.

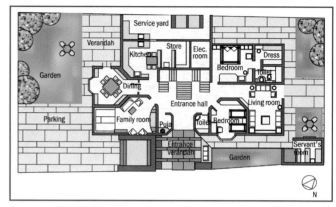

The ground-floor plan

Insulation

In this building, an attempt has been made to reduce the annual space-cooling requirement by nearly 30% with high level of insulation. The air-conditioning load was reduced from 36 to 26 tonnes of air-conditioning. Fly ash bricks, were used for masonry work. At every stage of the construction, plasticizer is used to decrease voids in the mortar. All the terraces and sloping roofs are finished with white China mosaic and insulated with 50-mm thick expanded polystyrene and sand cement plaster. All the external walls are insulated with 40-mm thick high-density thermocole. These thermocole sheets are adhered over the first coat of plaster by an adhesive (bitumen). Over these sheets, galvanized expanded wiremesh was anchored before application of the second coat of sand-faced plaster. All the walls are externally finished with white paint, which reflects most of the sunlight.

Construction details (L to R) for insulation, waterproofing, air-conditioning ducts, service ducts underside the shading elements, window frames and electrical box protection

Shading-cum-service ducts

1.2-m wide projections all around the building also work as service ducts to carry all the utility services like electricity, water supply, fan coil units of air-conditioning, etc., in addition to shading the walls. These projections are finished at the top by white China mosaic and covered at the bottom by aluminium false ceiling, which can be opened.

Indoor air quality

A major issue with a super-insulated, air-tight house is the quality of the internal air. In a poorly designed system, the low air change rate can lead to persistent odours, high humidity level, and even fungal growth. Problem of light and air quality is eliminated in this house by provision of north light and fresh air unit at the top of entrance hall. This acts as a light shaft, and also aids in the ventilation of the house.

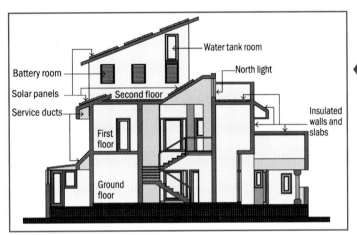

A section of the building showing integration of solar photovoltaic panels, location of battery room, service ducts, north light, and insulated walls

Solar photovoltaic and solar water heating systems

The solar PV (photovoltaic) system is both technically and economically suitable for household electrification in the long run. Once the decision was made to design a house with PV system, two years of electricity bills from the client's existing residence were studied to work out estimated consumption of energy per day. It was concluded that a connected load of about 18 kW is required to fulfil the client's need.

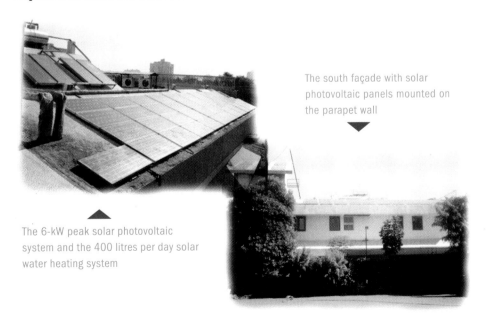

The 6-kW peak solar photovoltaic system and the 400 litres per day solar water heating system

The south façade with solar photovoltaic panels mounted on the parapet wall

The installed PV system has a battery bank capacity of 600 Ah/120 V nominal. The charging is accomplished during daylight hours by the 6-kW$_p$ solar modules. The charging of battery bank is controlled by power conditioning unit to prevent excessive discharge or overcharge.

The hot water requirement of the building is met by a 400-litre-per-day solar water heating system.

Building automation system

A building automation system is integrated with the electrical system to reduce energy wastage. The system hardware consists of various input sensors such as

- occupancy sensors and temperature sensors for overall utility management
- lux sensor for measuring ambient light levels
- water sensors for monitoring and controlling underground and overhead tanks' water levels
- breakglass, smoke detectors, and magnetic contacts for security applications.

The inputs from various sensors are processed by the PLC (programmable logic controller) and it controls their allied output systems according to the programme fed into it. The outputs that are controlled by the PLC are light fixtures, fans, 5-amp plugs, pressure pump, softening pump, air-conditioning, and security and fire-alarm system.

The hardware is fully interactive in nature—inputs from a particular sensor will not only activate its directly related output device, but also other relevant output systems.

A project of this nature not only extends the architect's responsibility to perform as a technocrat, but also supports newly developed systems. Mahendra Patel's house is a good example of a functional solar house with automation system.

At a glance

Project details

Building/project name Residence for Mahendra Patel
Site address 15, Kairvi Society, Bodakdev, Ahmedabad
Building type Residential
Climatic zone Hot and dry
Architect Pravin Patel
Year of start/completion 1996/97
Client/owner Mahendra Patel
Built-up area 550 m²
Cost of the project Rs 21 million (This includes construction work, finishes, solar systems, electrical works, security systems, air-conditioning systems, and interior work)

Design features

- A connected load of 18 kW is used to fulfil the client's need without compromising on any comfort
- Air-conditioning load reduced from 36 to 26 tonnes by passive solar interventions
- Fly ash bricks are used for masonry work
- External walls and roof are insulated
- Windows are with double glass shutters
- Walls are finished with white paint, which reflects heat
- There are 1.2 metre projections all round the building that work as service ducts to carry all the utility services like electricity, water supply, fan coil units for air-conditioning, and also as a shading device
- Problem of air and light quality is eliminated by provision of north light and fresh air unit at the top of the entrance hall
- Solar photovoltaic and solar water heating systems are used
- Building automation system is used to minimize energy wastage

Temperature check for Ahmedabad

(see Appendix IV)

Cold	Cool	Comfortable	Warm	Hot	
		●			January
		●			February
			●		March
			●		April
				●	May
				●	June
			●		July
			●		August
			●		September
			●		October
		●			November
		●			December

Add the checks

		4	6	2

Multiply by 8.33% for % of year

Comfortable	33
Cooling	67

Solar passive hostel, Jodhpur

Architect Vinod Gupta

A solar passive building for the hot and dry climate

Editor's remarks

Built in the hot and dry climate of Jodhpur, this building uses massive structure, insulation, proper orientation, and a wind tower for moderating internal diurnal temperature variation. Water shortage prevented use of an elaborate evaporative cooling system, which would have been very effective.

The solar hostel was put up as part of the research project undertaken by the Centre of Energy Studies, Indian Institute of Technology, Delhi. Although energy conservation was stated as the objective, the design attempted to test and demonstrate suitable methods of providing thermal comfort in the hot and dry climate of Rajasthan.

A view of the solar passive hostel

Design, materials, and techniques

The building had to be designed to house 14 double room suites for married students. The two-storeyed building has seven suites on the ground floor and seven suites on the first floor. Each suite is provided with a toilet (about 4 m² floor area), one lobby, and a small courtyard. The ground floor that has seven double rooms is partially sunk into the ground to take advantage of the earth's thermal storage and insulation effect. The wind tower, erected over the lobby of the first floor, is connected to the ground floor through the staircase and supplies cool air to the seven units. The hot room air exits by means of smaller chimneys over each room.

The protection of the roof and its treatment is important because it is a major source of summer heat gain. The roof has been insulated by providing small inverted terracotta pots over the stone slabs and filling up the intervening spaces with lime concrete. Stone masonry has been used for walls because it is a local material and can provide good thermal mass to balance out diurnal temperature variations. The wall thickness varies from 0.30 m to 0.45 m. Wind tower helps to ventilate the heat out of the room during late evenings and nights.

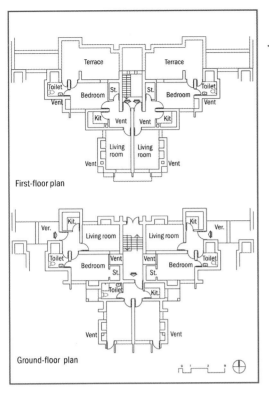

Plan for the solar passive hostel, Jodhpur

The design of the building has a set of rooms partially underground. This did not pose any major construction problems in Jodhpur because of its low water level. Toilets were also placed at the same level as these can be drained to the sewage system or decomposition tank/pit as the case may be. The partially underground configuration has a moderating influence on the temperatures, reducing the solar heat gain on the walls and cooling like a basement, which are also traditionally built.

For this building, an improved design of wind tower was made with built-in evaporative cooling to lower temperatures. Higher airflow rates and the evaporative capacity of the new wind tower can be fully utilized at night in summer to cool the building mass to lower temperatures.

Passive solar features

The air being very dry, evaporative cooling in summer can prove to be very effective in Jodhpur. Unfortunately, as water is a scarce commodity in summer in Jodhpur, any system that depends upon water for cooling is bound to fail. The design, therefore, uses a favourable orientation, a massive structure, and air gap in the roof for insulation, reflective external finishes, deep sunshades, and finally a wind tower for making use of the cool winds. An experimental evaporative cooling system using wires for water distribution has also been installed on the wind tower.

Wind tower

The prevailing direction for cool winds in Jodhpur is the south-western. Window apertures are difficult to provide in this orientation, as it is the least favourable from the point of view of solar radiation. To overcome this problem, a wind tower concept was used. The tower facing the wind direction has been located over the staircase, thus minimizing costs. Cool air is provided to each room from this tower and normal windows or smaller shafts (towers) facing the lee of the wind have been provided to distribute the cool air throughout the building. The tower catches only the cool wind from the south-west, avoiding warmer air from other directions.

A view of the wind tower and solar water heating system of the hostel building

Roof insulation

The commonly used building material in Jodhpur is the local stone. Blocks of this light-coloured stone have been used for walls in the building. Large slabs of stone have been used for roofing, staircases, partitions, and lintels over windows. The roof has been insulated by providing small inverted terracotta pots over the stone slabs and filling up the intervening spaces with lime concrete. Since very few manufactured materials have been used, this is a low embodied energy building.

Window design

Since the university is normally on vacation during the worst summer months, winter comfort is as important as summer comfort. South-facing windows have been provided in most of the rooms. To prevent heat loss during night, solid timber shutters have been provided in addition to glass.

Performance The monitoring results in various rooms of the hostel show that internal temperature remain nearly constant round the day without much fluctuations on a typical summer and winter day, while the ambient temperature has large diurnal variation.

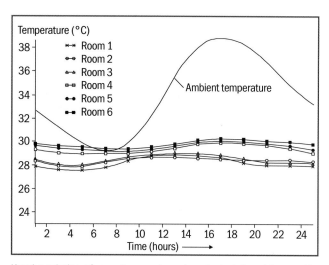

Hourly variation of room temperatures for a typical day of the month of June

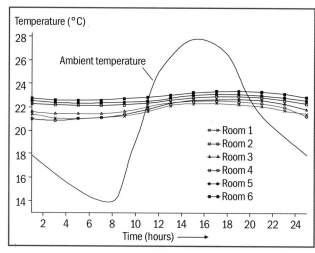

Hourly variation of room temperatures for a typical day of the month of February

At a glance

Project details

Project description Hostel containing 14 suites for married students, located at Jodhpur
Architect Vinod Gupta
Energy consultants N K Bansal and M S Sodha
Project completion 1984
Client/owner University of Jodhpur
Installed renewable energy system A common bank of solar water heaters has been installed to meet all hot water requirements.

Performance results are available with Prof. M L Mathur, University of Jodhpur, Jodhpur and Dr N K Bansal, Professor, Centre of Energy Studies, Indian Institute of Technology, New Delhi

Solar passive hostel, Jodhpur

> ### Design features
> - Favourable orientation
> - Low embodied energy by making minimal use of manufactured materials
> - Heavy construction to balance out diurnal temperature variation
> - Roof insulation by air gap
> - Light-coloured building to reflect heat
> - Wind tower with evaporative cooling for summer cooling
> - Building partially sunken to moderate internal temperature
> - South-facing window with deep sunshades to cut off summer sun and to let in winter sun
> - Solid timber shutters in addition to glass shutters to prevent heat loss during winter nights.

Temperature check for Jodhpur

(see Appendix IV)

Cold	Cool	Comfortable	Warm	Hot	
	●				January
		●			February
		●			March
			●		April
				●	May
				●	June
			●		July
			●		August
			●		September
			●		October
		●			November
		●			December

Add the checks

	1	4	5	2

Multiply by 8.33% for % of year

Heating	8
Comfortable	33
Cooling	58

Climatic zone: moderate

Residence for Mary Mathew, Bangalore

TERI office building-cum-guest house, Bangalore

Residence for Mary Mathew, Bangalore

Architects Nisha Mathew and Soumitro Ghosh

This residence-cum-office building in the moderate climate of Bangalore uses solar energy to optimize comfort and heat water

Editor's remarks

At a time when human relationship with the ground and sky is cut off by multi-storeyed high-rise energy guzzlers, the Mathew house makes a case for the urban house with a traditional garden court, determined by limitations of space, affordability, and climate.

The city of Bangalore is located in the moderate climatic zone with annual mean maximum temperature of 28.8 °C, annual mean minimum temperature of 18.4 °C, and the annual range of mean temperature being about 10.4 °C. The relative humidity varies between 30% and 80%. Climatic conditions being generally within a favourable range, the building designs do not require any special interventions to provide thermal comfort. Provision for reduction of direct solar gain and heat transfer to interior, and increase in heat loss by ventilation helps in achieving thermal comfort conditions. The south and south-west sides are protected and northward orientation of the bedrooms is favourable.

Determined by the constraints of space, ways of life, and affordability, the Mathew house makes a case for a urban house with a traditional garden court. The house abuts the road on its 9.15-m front and goes in 26-m deep.

▲ The southern façade with minimal openings and housing the service areas provide a thermal barrier against heat gain

Planning as a climatic response

The architects have attempted a formal response to the climate by the creation of a 'thick wall' along the south-west; thus a block is perceived along the main road approach to the house. The idea was to create a conceptual 'wall' on the southern/south-western side, which comprised largely the masonry surface within which services such as toilets, pantry, kitchen work space, and servants' room were located. The depth of the south-west wall was used to shield the heat and provide pockets for openings located on this 'wall' to pull in south-west breeze. The courtyards were located in the north-east and north-west to provide a comfortable outdoor living space.

Only as one moves through the the house is the beginning of a perforated configuration realized and the sequential arrangement of spaces revealed. The emphasis on the environment as a backdrop is evident in the treatment of spaces. The 'verandah' and the garden court form focal points around which the interior spaces revolve. The garden court is formally defined by the water tank's pivotal position at its corner. The south-west wall flanks the verandah, which in differing 'densities' encloses the service spaces and shields the garden court from the sun.

The private chambers are accessed through a sequence of galleries on the ground and the upper level as transition. The morning sunlight penetrates each of these spaces. Open spaces of a small scale are attached to the upper chambers and extended from the study as well.

Low embodied energy

As an effort towards reducing embodied energy, the architect has used low-cost, low-energy construction methods to complement the austerity of the spaces. The structural system of load-bearing walls and jack arch roof slab with reinforced precast area with minimal reinforcement was used. It became a guide for the articulation of spaces, fenestration, interior surface, and exterior façade with a bay width of 1.18 m.

1 Terra cotta tile flooring
2 Half weatherproof tile (WPT) coping
3 Floor base
4 Concrete jack arch slab
5 Precast concrete jack arch slab beam
6 Precast WPT jack arch panel
7 Edge waterproofing
8 Plaster drip mould
9 Single dressed granite lintel
10 Groove pointed wire cut brick
11 Plastered window cill
12 In situ gray cement skirting
13 Earth
14 Concrete plinth Beam
15 Granite rubble foundation

▲ Innovative construction methods for walls, roof, and foundation provide energy and resource efficiency

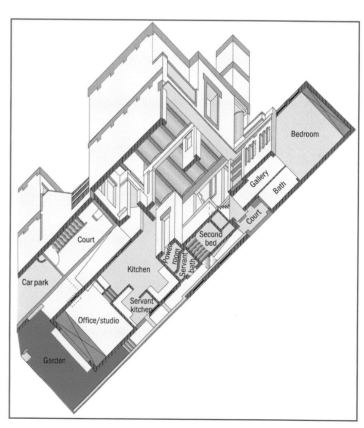

Axonometric view showing ▶ progressive construction

Roof insulation

Roof insulation was provided by using a roof system of precast hollow terracotta curved panels with nominal G I reinforcement. A nominal layer of concrete of only 2-inch thickness at the crown of panel was poured into place. The hollow terracotta layer works as heat-resisting layer.

Daylighting

Natural lighting is extensively used in the north-east and north-west by hollowing out courtyards, which become permanent sources of light and ventilation. Slabs made of local stone in place of concrete fins provided the slit windows.

▲ View of well daylit dining, living, and pantry areas

White is the primary colour of the interior and exterior spaces revealing textures of rough, rugged, and smooth surfaces and adding to heat reflectivity of the wall surfaces. The house is characterized by the architects' vivid articulation of spaces and the exploration of the garden court theme.

Installed renewable energy systems

Solar water heating panels and back-up geysers are used almost throughout the year and is the only source of water heating.

Performance

- No air-conditioners are being used in the house.
- Natural ventilation, solar access, and natural lighting are good.
- Electricity consumption in summer is approximately 240 units per month. In monsoon, the consumption is approximately 255 units (kWh) per month and in winter, it is approximately 160 units per month. This power consumption is inclusive of pumping water for garden (twice a day in summer) and house from borewell, in view of absence of corporation water supply for the home and office. However, the consumption figures showed marginal increase after computerization of the office.
- None of the lights are used during daytime other than during cloudy monsoon days. Maximum artificial light used is 2.4 kWh per day, of which daily night peak usage is about 0.9 kWh–1.1 kWh, which is a bit high due to the location of the house in isolated area and lack of streetlighting. Lights are not used before 5.00–5.30 p.m. in winter and 6.00–7.00 p.m. in summer.

▲ View of gallery to master bedroom. Stone chapti slit windows provide natural lighting

At a glance

Project details

Building/project name Residence for Mary Mathew
Site address 2 Temple Trees Row, Viveknagar Post Bangalore – 560 047
Building type Residence-cum-office
Climate Moderate
Architects Nisha Mathew and Soumitro Ghosh
Year of start/completion August 1995 to June 1996
Site area 237 m²
Ground floor area 149 m²
First floor area 87 m²
Total floor area 236 m²
Total cost Rs 1.1 million

Design features

- Natural lighting is extensively used in the north-east and north-west by hollowing out courtyards, which become permanent sources of light and ventilation.
- Roof insulation was provided by using a roof system of precast hollow terracotta curved panels with nominal G I reinforcement. A nominal layer of concrete of only 2-inch thick at the crown of panel was poured into place. The hollow terracotta layer works as heat-resisting layer.
- A thick 'wall' on the southern/south-western side, which comprised largely masonry surface within which were located the services such as toilets, pantry, kitchen work space, and servants' room. The depth of the south-west wall was used to shield the heat and provide pockets for openings located on this 'wall' to pull in south-west breeze.

Temperature check for Bangalore

(see Appendix IV)

Cold	Cool	Comfortable	Warm	Hot	
		•			January
		•			February
		•			March
			•		April
			•		May
		•			June
		•			July
		•			August
		•			September
		•			October
		•			November
		•			December

Add the checks

		10	2	

Multiply by 8.33% for % of year

Comfortable	83
Cooling	17

TERI office building-cum-guest house, Bangalore

Architects Sanjay Mohe and V Tushar

An energy-efficient, eco-friendly office building with minimal environmental footprint

Editor's remarks

The proposed TERI office building-cum-guest house has been designed in response to various site constraints. Innovative use of on-site sources and sinks in this building would set an example for many such buildings of the future.

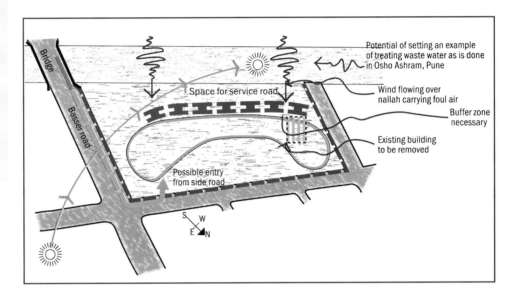

▲ Site analysis showing evolution of design in response to site conditions

The site and the requirements

This project is designed to house an office block with approximately 75 workstations and a small guest house attached to it. The nature of work of TERI (Tata Energy Research Institute) personnel demands interaction spaces, conference rooms, library, laboratory, etc. The dining hall and recreation area are shared by the office and the guest house.

The site is located at Domlur, about 3 km from the Bangalore airport. It is a long and narrow site with roads on the eastern and northern sides. The western side has an open ground and the southern side has an open drain about 9 m wide. This drain, with its foul smell, dictates the design development, as wind comes from the south, bringing in the foul smell into the site.

Zoning

Entry to the building is from the road on the northern side, which is less busier, as compared to the one on the east. The office block is kept towards the east, close to the main road for high visibility and the guest house is located towards the quieter western side. The open space between the office and the guest house can be used for future expansion of either the office block or the guest house.

Design response

Reclamation of nallah

The first reaction to the design problem was to improve the condition of the drain and make it a pleasant landscaped element on the line of Nallah Park, next to the Osho Ashram in Pune. This would be done by using plants that absorb impurities as well as with the help of basic filtration and aeration. This would be a major civic project and would involve undertaking a longer stretch of the *nallah*.

Though this would be an ideal long-term solution, the architects had to respond to the present site conditions and design a building, which can eventually open up towards the drain (when it turns clean).

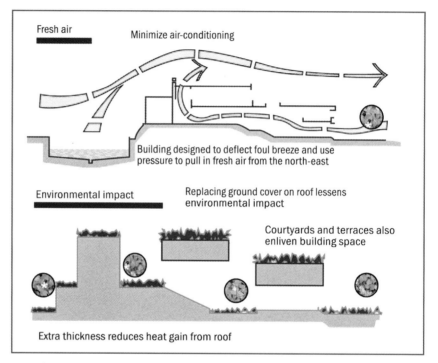

▲ Form developed to naturally ventilate the entire building thus minimizing air-conditioning requirements. The section has been developed to tackle the negative impacts of the *nallah*

Passive ventilation techniques

The building opens towards the northern side, taking advantage of glare-free light. The wall towards the south (*nallah* side) is made into a blank wall, allowing the breeze to flow over the building, which, in turn, creates negative pressure and starts pulling fresh air from the north into the building. The sections are worked out in a way to allow hot air to rise towards the top and make the building breathe. The south wall is made into a double wall, firstly to provide insulation from the southern sun, and secondly, to heat up the void between the two walls creating negative pressure, thereby enhancing the convection currents. Additional earth berms are created towards the drain side as a buffer.

The sections are naturally ventilated with the air flowing from the ground floor to the terrace because of the open nature of the volumes. The ventilation is also enhanced by the use of solar chimneys and vents. These are effective on both sunny and windy days.

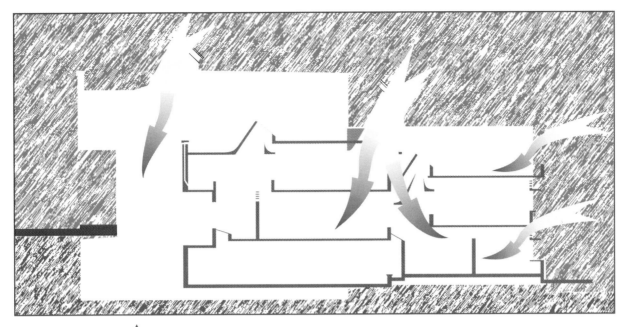

▲ The roof lighting system provides uniform daylighting throughout the building. The roof lighting panels are proposed to be of semi-transparent solar photovoltaic panels

Daylighting design

There was a detailed daylighting study and the fenestrations have been designed so that requirement of artificial lighting is minimal during daytime. By creating atrium spaces with skylights, the section of the building is designed in such a way that natural daylight enters into the heart of the building, considerably reducing the dependence on artificial lighting. Also the skylight roof is proposed to be made of semi-transparent solar photovoltaic panels. Energy-efficient lighting using efficient lamps, luminaires, and control strategies have been planned.

Rainwater harvesting

A scheme of rainwater harvesting, which would be used to water the plants, has been worked out. Water run-off from the roofs and from the paved area will be collected at various levels in small open tanks on the terraces and in a collection sump below. This water will be used for landscaping.

Section explaining natural ventilation and rainwater harvesting in the roof ▶

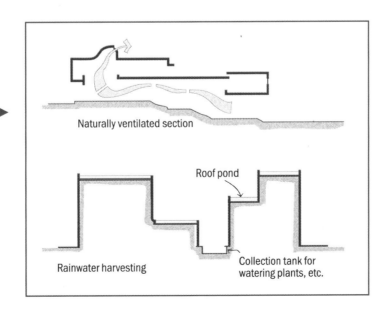

Roof garden

The ground cover, which is disturbed due to the building of this structure, will be replaced on the rooftop, in the form of terrace gardens, giving insulation to the building and reducing solar radiation. The ground-covered roof provides good thermal insulation and moderates fluctuations in temperature.

Renewable energy systems

A 5-kW peak solar photovoltaic system has been planned, which would be integrated with the roof skylights. The photovoltaic roof would provide daylighting and generate electricity as well. A solar water heating system would meet hot water requirements of the kitchen and the guest rooms. The other features planned in the building are an effective waste and water management system, a centralized uninterrupted power supply, and a cooler kitchen that seek to reduce internal heat. The materials come together to form a building with low embodied energy.

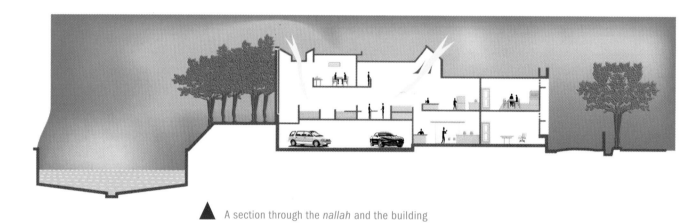

▲ A section through the *nallah* and the building

Thus, the proposal addresses not only thermal comfort but also visual appeal and environmental issues. An integrated approach to building and system design has been adopted.
The design conveys a dextrous interplay of natural elements with the built form, which reduces energy demand at end-use. Use of natural elements is exemplified by:
- **sun** for water heating, electricity generation, creating stack effect by use of solar chimneys, etc.;
- **sky** for daylight and also as heat sink;
- **air** to create convection currents within the building through wind-induced vents, use of venturi effect;
- **earth** for roof gardens and earth berms for insulation; and
- **water** for rainwater harvesting, roof ponds, and fountains for humidification.

There is an attempt to create this building as a test model, which will demonstrate the conservation of energy through post-occupancy monitoring and further develop and perfect the system.

At a glance

Project details

Site Domlur Stage II, Bangalore
Client Tata Energy Research Institute, Bangalore
Climate Moderate
Architects Sanjay Mohe and V Tushar
Year of onset 1998
Built-up area 2450 m²
Cost Civil, sanitary, and plumbing costs tendered at Rs 21.9 million
Consultants SEMAC and TERI (PV and other systems)
Status Ongoing

Design features

- Innovative treatment of south wall to cut off foul smell from adjoining *nallah* and induce ventilation
- Naturally ventilated building reduces air-conditioning requirements
- Roof gardens as roof insulation
- Rainwater harvesting
- Daylighting and energy-efficient lighting
- Building-integrated solar photovoltaic panels to generate electricity
- Solar water heating for hot water generation
- Provision to treat *nallah* in the future

Temperature check for Bangalore
(see Appendix IV)

Cold	Cool	Comfortable	Warm	Hot	
		●			January
		●			February
		●			March
			●		April
			●		May
		●			June
		●			July
		●			August
		●			September
		●			October
		●			November
		●			December

Add the checks

		10	2	

Multiply by 8.33% for % of year

Comfortable	83
Cooling	17

Climatic zone: warm and humid

Nisha's Play School, Goa

Office building of the West Bengal Renewable Energy Development Agency, Kolkata

Office-cum-laboratory for the West Bengal Pollution Control Board, Kolkata

Silent Valley, Kalasa

Vikas Apartments, Auroville

La Cuisine Solaire, Auroville

Kindergarten School, Auroville

Visitors' Centre, Auroville

Computer Maintenance Corporation House, Mumbai

Nisha's Play School, Goa

Architect Gerard de Cunha

An innovative school design with low embodied energy and maximized use of natural ventilation and daylighting

Editor's remarks

The Nisha's Play School, located in the warm humid climate of Goa, uses innovative building design and detailing to induce maximum natural ventilation and daylighting. Use of innovative materials and methods of construction has reduced the overall energy content of the building.

Nisha's Play School practises the playway method of education. In this method, each classroom had to have spaces for formal teaching, informal work, and a separate dolls house where children interact with each other.

▲ A view of the front façade. The building opens up as it stretches behind to follow the natural downward slope

Site planning and building design

The plot has a frontage of 20 m and a depth of 40 m (area 800 m²) and slopes down from the road at an angle of 30 degrees till half the depth and is then flat. It is located in a valley at the edge of the forest and has a lot of tree cover, making the site rather dark and poorly ventilated. It falls into the 'warm humid climatic zone' and the trees offer shelter from solar radiation to a great degree, but retard the movement of air and cut off light considerably. The introduction of natural light and the need to induce natural ventilation (preventing a build-up of humidity) were the two prime determinants in the architectural design.

Site planning

As the flat area was completely covered by trees, it was left as the playground and it was decided to build on the slope exactly balancing cut and fill. It meant accepting the fact that little children would have to walk up and down. The side set backs used were 3 m on the west and 5 m on the east, which helped to retain all the trees.

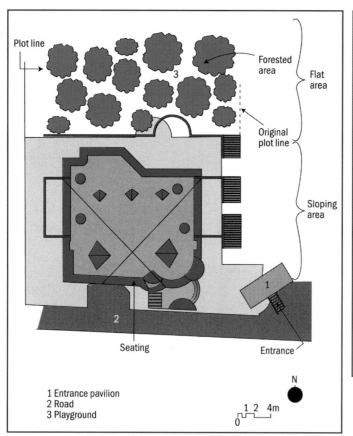

The site plan of the building. The irregular shaped features on the roof mark position of the skylights for daylighting

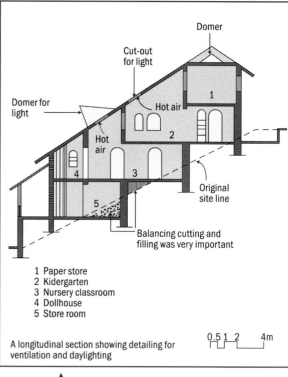

Section of the school. Floors have been staggered so that each level has contact with the exterior roof from which light is brought in from cut-outs or dormers and air is also naturally extracted through these openings

Building form

Besides building on a slope and the need to keep trees, the intake of natural light into the building, inducing breeze and the need to maximize the usage of space greatly determined the final form. It is within this rigid form that the architect went about creating variety and spatial experience. The building revolves around a circular core, which has the stairs, the chute, and the bell.

Roof

The roof has five cut-outs and five dormers (refer roof plan) for the purpose of day lighting and ventilation. The induced ventilation aspect of the design has been most successful and fans are hardly used. Around Christmas, some of the teachers wear sweaters, which seems very strange for Goa. Using a filler slab with the filler being thermocole packing or empty bottles has reduced the heat intake from the RCC (reinforced cement concrete) roof. Perforated concrete *jalis* below roof slabs also help in cooling the rooms by inducing ventilation. There are many air intakes at floor level ensuring that the whole building gets ventilated.

Embodied energy

Another aspect was the low level of energy consumed in the construction of the building, with natural stone as an important ingredient of the building. All windows and doors have been recycled and all forms of plaster eliminated. Waste tiles have been used for many floors. Filler slabs have been used for roof construction.

The toddlers' class with filler slab roof (thermocole and liquor bottles area used as fillers). The roof cut-outs admit daylight

- Thermocole packing filler
- Empty bottle filler
- Cut-out for light
- Lion mural
- Toys for kids to help themselves

Viewed in its basic form, the building derives its design from climatic considerations, but within this framework, an attempt has been made to have each classroom as one of a kind, which stimulates the creativity, and imagination of children. Elements such as window grills are being used as teaching aids. Children in Nisha's Play School tend to come in early and leave late, part of it being due to the friendliness of its architecture. The materials and construction techniques used in the building are given below.

- *Structure and building material*   Load-bearing structure with vertical and horizontal RCC bands for earthquake protection
- *Foundation/retaining walls*   Random rubble laterite masonry in cement mortar
- *Superstructure*   Exposed 9-inch brickwork to maximize space and random rubble masonry where space was not a constraint

- *Roofing/intermediate floors*   RCC with various fillers—bottles, thermocole packaging, etc.
- *Doors and windows*   Mostly recycled old doors and windows, steel grills, brick jalis
- *Flooring*   Red oxide with inlay in black in classrooms / circulation space. China mosaic in doll houses and open area
- *Dado's/toilet floors*   China mosaic waste with inlay of waste mirrors.
- *Shelving*   Polished Cadappa embedded in brickwall

At a glance

Project details

Name of the project Nisha's Play School
Address Nisha's Play School, Torda, Savador Do Mundo, Bardez, Goa
Climate Warm and humid
Design team Gerard Da Cunha, Annabel Mascarenhas, Lisa Thomas, Nirmala D'Mello
Structural consultant Madhav Kamat and Associates
Area of project 480 m²
Cost of project Rs 1.6 million
Year of completion January 1997

Design features

- The building design and form evolved out of demand to maximize daylighting and induce natural ventilation
- Use of locally available materials, waste materials, and materials with low embodied energy.

Office building of the West Bengal Renewable Energy Development Agency, Kolkata

Architect Gherzi Eastern Ltd

An integrated approach to building design to achieve energy efficiency

> **Editor's remarks**
>
> This office building showcases passive solar architectural principles for warm and humid climate. Well-lit and naturally ventilated round the year, this building also boasts of a 25-kW peak grid interactive solar photovoltaic system.

The enormous amounts of energy consumed in office buildings are a cause for great concern. For this reason, the WBREDA (West Bengal Renewable Energy Development Agency) decided to incorporate energy efficiency measures in the building design and to use appropriate non-conventional energy systems. To facilitate such energy efficiency measures, the building has been designed using the basic concepts of solar architecture. The building layout, internal planning, and selection of material have been carefully considered in order to reduce energy consumption.

▲ The south façade with light shelves for daylighting. The west façade is devoid of any fenestration to prevent direct gains.

The concept

The site is a rectangular plot longer in the east-west direction. Kolkata being in the warm humid zone, the evolution of the design was based on the following determinants.

- East, west, and south side must be protected from direct sun.
- Ground surface must not reflect and radiate heat.
- In warm and humid climate, natural ventilation must be of highest priority.
- There should be ample provision of cross-ventilation.
- Natural elements like vegetation and water must be utilized to moderate microclimate.

Design strategies

A study of Kolkata's climate shows that from April to September during the year, mechanical cooling of the building is required in order to provide thermal comfort. Various strategies have been incorporated, as outlined below, to cool the building, where an effort has been made to use the on-site heat sources and sinks to allow the heat exchange between the building and the surroundings. The energy consumption would be drastically reduced due to these design strategies.

The building has been formulated as a rectangular structure of 26.7 × 17.8 m. The total built-up area is 2026 m² with ground plus three stories of office spaces. In addition to the office spaces, the office building has provision for exhibition, conference, library, and documentation. The basic design of the building has been conceived with following main features.

Evolution concept of building design ▶

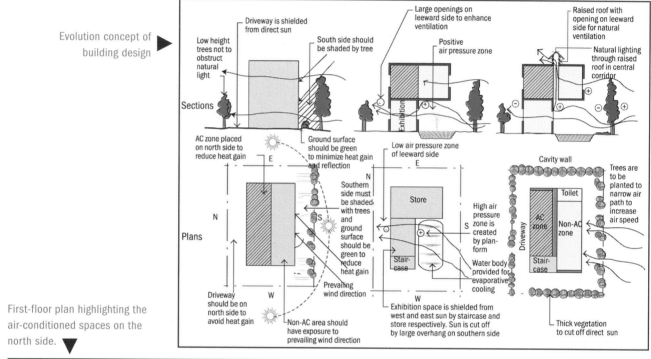

First-floor plan highlighting the air-conditioned spaces on the north side. ▼

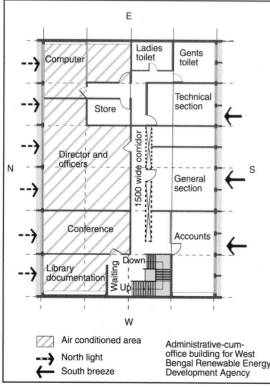

Space planning

The form of building plan has been conceived as a rectangle. Shorter sides of the rectangle are facing east and west to reduce heat gain. Air-conditioned areas are kept on northern side of the building. The walls on east and west face are devoid of fenestration. The areas, which are not used frequently, like store, staircase, and toilets, are placed on the eastern and western sides as buffer against direct solar heat.

Landscaping

The ground that faces the southern and eastern side of the building would be covered with grass or water to minimize heat gain from the surroundings. Use of vegetation and water bodies has been encouraged to modify the microclimate.

Ventilation

Non-air-conditioned areas are located on the southern side to take advantage of the prevailing wind during hot and humid period. For better cross-ventilation, a portion of the roof has been raised and used as solar chimney. Southern face of the wall has been protected by overhangs. These overhangs have been designed to function as light shelves for even distribution of daylight.

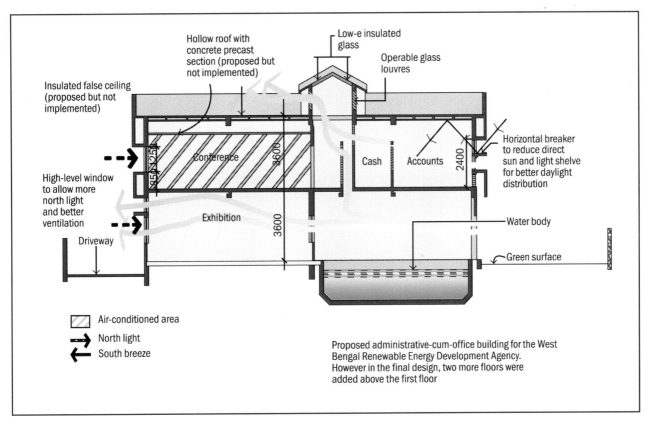

▲ Section of the WBREDA office building showing ventilation strategies

The building is virtually divided into two parts: north and south blocks serviced by a corridor in between. To best utilize the prevailing south breeze, a water body has been created in the southern part of the building at the ground level, which has, in turn, increased the aesthetic value of the concept. The south breeze blowing over the water body gets trapped at the bottom of the building and the same is vented through the building with suitable cut-outs, ventilators, windows, so that the cool south breeze can blow up to the deepest portion of the building. The calculated opening provided act as vent shaft taking care of the cross-flow of south wind and taking out the hot air from the non-air-conditioned areas. The raised roof covered by low e-glass acts as a solar chimney and creates draft for the ventilation of the spaces.

Daylight integration and energy-efficient lighting design

The basic requirement of any lighting installation is to provide sufficient light in the right place at the required time.

The lighting scheme for the WBREDA office building has been designed to provide the desired quality, and the recommended quantity of light as per ISI standard. While providing a good colour rendition (cool daylight appearance), illumination levels provided in different areas of the proposed office are as follows.
- Entrance hall and reception area – 150 Lux
- Conference room and private offices – 300 Lux
- Open plan office – 300 Lux
- Stairs –100 Lux
- Corridor – 70 Lux

The following lighting equipment were recommended for the office building
- Linear fluorescent lamps (36 W high efficacy tubelights) to illuminate the interior work zones
- Reception areas, entrance and elevator lobbies, corridors, staircase, and other public areas to be fitted with compact fluorescent lamps
- Mirror optic luminaires
- Good quality ballasts for control of all discharge lamps

Daylighting

The windows provided on the north and south façades are adequately sized to provide sufficient daylight for most of the day and throughout the year, except during the monsoon period. The sizes of shading devices in the form of fixed louvres (overhangs, side fins) have been optimized on the basis of solar geometry. The entry of daylight into the work areas is enhanced by providing a light-shelf in the south windows.

All the circulation spaces like staircase, lobby, and corridors are naturally lighted by way of raised roofing.

Lighting control

Recommended lighting control strategies for the WBREDA building are
- time-based control,
- occupancy-linked control,
- daylight linked control, and
- localized switching.

These controls are yet to be installed in the building.

Insulation

Budgetary constraints prevented use of insulation in the building. However, conduction heat gains would have been reduced by increasing the thermal resistance of the envelope, mainly by the use of thermal insulation in the roof and walls. The insulated walls and roof not only lower the energy consumption but also allow lower mean radiant temperature (effective temperature of room surfaces) to provide better thermal comfort. The insulation would have reduced the air-conditioning load by 40%.

Glazing system

Air-tight double-glazed windows in the air-conditioned areas and single-glazed windows for non-air- conditioned areas were recommended to check the unwanted heat gains through infiltration throughout the building, and through conduction in the air-conditioned areas. Double-glazed windows were expected to reduce the mechanical cooling load by about 12%. However, double-glazed windows could not be provided due to lack of funds.

The raised roof acting as a solar chimney has a low e-glazing system to reduce heat gains.

Use of renewable energy system

The client has assured that there would be no hot water demand in the building. Although the solar hot water systems would be installed outside the office for demonstration purposes, these systems have not been considered for integration into the building.

At present, a 25-kW roof-mounted grid-interactive solar photovoltaic system meets major part of the building electrical load.

▲ The roof of the solar chimney is of low e-glass. A 25-kW peak solar photovoltaic system has been installed on the roof

Test beds for experiment on renewable energy systems

A test bed has been provided at the south side of the plot in the open space of an approximate area of 167 m². All vehicular movements have been restricted by providing the drive-way in the north and east parts of the building, while the south part is kept free for the demonstration of renewable energy systems and other R&D activities of the WBREDA.

Performance The building has just been occupied and no monitoring results are available. However, the architect observes that the building is well lit during daytime and does not require artificial illumination. The ventilation strategy also works well, says Mr Pramanik, one of the principal architects for the project.

At a glance

Project details

Building type Commercial (Office building)
Location Kolkata
Climate Warm and humid
Architect Gherzi Eastern Ltd
Energy consultant TERI (Tata Energy Research Institute), New Delhi
Year of completion 2000
Client/owner West Bengal Renewable energy Development Agency
Plot area 10 895 m²
Built-up area 2026 m²
Total project cost Rs 16.3 million, excluding the cost of solar photovoltaic system and air-conditioning

Design features

- Space planning done so as to reduce air-conditioning loads.
- Ground surface facing southern and eastern sides of the building to be covered with grass.
- Use of vegetation and water bodies to be encouraged as a modifier of microclimate.
- Office spaces naturally lit by way of raised roofing with low e-glass and light shelves.
- Proper design of shading device to cut off direct gains and let in daylight.
- Removal of internal heat by incorporating ventilation device.
- Energy-efficient lighting with integration of daylighting.
- 25-kW$_p$ grid-interactive solar photovoltaic system for meeting major part of the building load

Temperature check for Kolkata

(see Appendix IV)

Cold	Cool	Comfortable	Warm	Hot	
		●			January
		●			February
			●		March
			●		April
			●		May
			●		June
			●		July
			●		August
			●		September
			●		October
		●			November
		●			December

Add the checks

		4	8	

Multiply by 8.33% for % of year

Comfortable	33
Cooling	66

Office-cum-laboratory for the West Bengal Pollution Control Board, Kolkata

Editor's remarks

This office-cum-laboratory building is an energy and resource-conscious architecture of eastern India. Efficient planning, proper fenestration, and shading design have brought about 40% savings over a conventional building of similar size and function.

Architects Ghosh and Bose & Associates

An office building in a tight urban setting that uses innovative planning and detailing to achieve energy efficiency

The building of the WBPCB (West Bengal Pollution Control Board) has put to use a number of technologies that aim to promote a more sustainable built environment. The coming years will show which of them we may adopt, refine, or perhaps even discard.

▲ A view of the building showing the north-oriented windows of office space and laboratories. The window sizes and shading devices have been worked out using scientific tools for fenestration design

Traditionally, the building and construction industry has been a major consumer of energy and natural resources. Hence, at the very outset, the WBPCB and the architects decided that, since the former was actively engaged in uplifting the environment, their office building should be an exemplary 'environment-friendly building'. The architects thus took it as a challenge to design an energy-efficient building applying various energy and resource-efficient building systems.

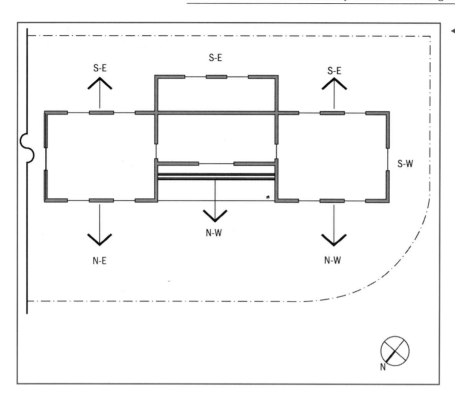

A conventional plan would have exposed large glazed area to south-east, north-east, and south-west, resulting in the unutilizable glare of direct sunlight and excessive heat gain

The functional requirements were broken down into three basic components.
- Fully air-conditioned laboratory wing of about 1115 m², with state-of-the-art laboratories
- A ventilated, non-air-conditioned office wing providing office space of about 1300 m²
- An ancillary wing housing entrance lobby, cafeteria, auditorium, training centre, library, and guest rooms.

Energy-saving features

The primary strategy of the architects was to achieve enhanced daylighting and optimum thermal condition of the building envelope, as the largest end-uses in office building are the lighting and air-conditioning systems. The first intervention to achieve this was judicious orientation of the building within site constraints.

Orientation

The WBPCB site was not suitable in this respect, being a long narrow plot facing north-west and south-east. A conventional plan would have exposed large glazed areas to south-east, north-west, and south-west, resulting in the useless glare of direct sunlight and excessive heat gain. By effective

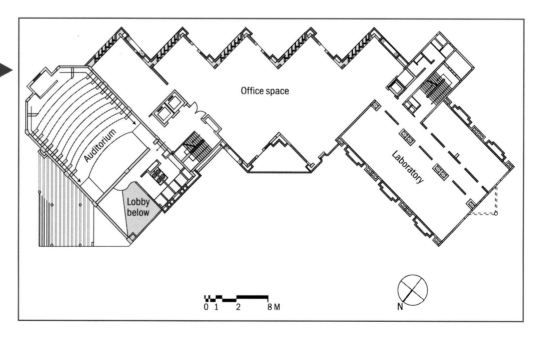

A typical floor plan showing the staggered plan-form to ensure maximum daylighting and ventilation and minimum direct solar gains

architectural design, the key laboratory and office spaces are oriented north–south for both daylighting, good ventilation, and optimum thermal condition.

Fenestration design

The shading devices, and window size and disposition vary according to the orientation of the walls. In addition, at the initial design stage, the architects ensured appropriate depth of the plan to maximize daylight penetration into the interior.

The highlights of the solar passive features are optimum window disposition and sizing to allow maximum daylighting, while minimizing adverse thermal effects. However, indiscriminate increase of glazing area to achieve this is counter-productive of causing glare and over-heating of the building. The glare from uncontrolled daylight necessitates the use of curtains and blinds with resultant increase in use of artificial lighting and cooling load.

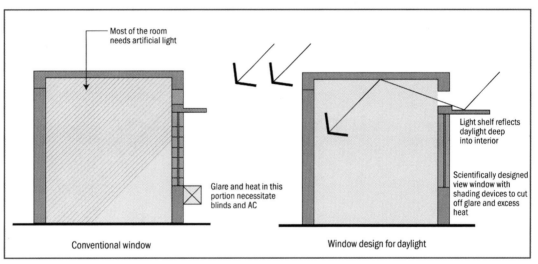

▲ Diagram showing a conventional window design and window designed for integration of daylighting and to cut off direct radiation

Design optimization

To arrive at an optimum solution, the entire interior was computer-simulated to test the light levels (at 1-m grid intervals) and thermal performance (with different window sizes). Window sizes were finalized after this exercise. Moreover, as the provision of windows in suitable orientation from the consideration of solar incursion and wind direction was important, maximum glazing was provided in the north-south direction, and minimum in east and west directions. This provides the advantage of solar heat in winter while minimizing it in summer. This orientation is also suitable for ventilation, in cities like Kolkata. This exercise had resulted in reduction of cooling load for air-conditioned laboratory area. The staircase and toilet blocks are located in that part of the building, which unavoidably faces west.

Scientific design of shading devices

The shading devices were designed specifically for different wall orientations to control the glare and reduce the thermal load on the building. For windows facing north, vertical louvres normal to the wall, capped by a horizontal member of the same width on top are adequate to provide the required shading. Windows facing south were generally shaded with horizontal louvres. Normally these horizontal louvres should extend much beyond the window, possibly to other windows at the same level, to avoid sunlight coming partly from the corners. Hence, instead of extending the horizontal members to any distance beyond the window on either side, two vertical louvres were provided at the two extremes. For windows oriented east and west, the recommended shading device is a combination of horizontal and vertical louvres. The horizontal louvre is normal to the wall but the vertical louvre is inclined at 30 degree towards the south, away from the normal to the wall. This has the advantage of letting in the winter sun during early mornings on the east façade and of completely cutting off the summer sun from morning to evening. The shades are designed so that the summer sun is cut off and the winter sun is allowed in. Again, computer simulations were done to test the efficiency of different shading devices—horizontal louvres, vertical louvres and a combination of the two. The predicted savings in energy consumption by doing this exercise has been summarized in Table 1.

Table 1 Consumption and savings on one typical floor

Area	Case	Lighting consumption (kWh)	HVAC consumption (kWh)	Total consumption (kWh)	Savings (%)
North-facing laboratory	Conventional window/shading	5070	11592	16662	–
	Final design	936	10080	11016	33.9
South-facing laboratory	Conventional window/shading	2150	5760	7910	–
	Final design	624	5328	5952	24.7
Office block	Conventional window/shading	12960	–	12960	–
	Final design	624	–	5160	24.7
One typical floor	Conventional window/shading	20180	17424	37608	–
	Final design	7170	15408	22578	39.8

Source TERI. 1996. **Design review of West Bengal Pollution Control Board Building at Salt Lake, Kolkata.** New Delhi: Tata Energy Research Institute [TERI report 1995RT65].

The consumption (lighting, HVAC, and total) is predicted annually assuming nine working hours per day and five working days per week.

The total saving (approximately 39.8%) is achieved by controlling window sizes/shading devices. Solar passive techniques with respect to orientation and depth of plan have led to a saving of about 2.6% over a conventionally designed building. Hence, due to a combination of correct orientation, depth of plan, window, and external shading, it is anticipated that the designed pollution control board building at Salt Lake will save approximately 41.5% energy annually over a conventionally designed building of the same size.

Energy-efficient lighting

Switching circuits and automation

To utilize fully the benefits of daylight in the interior, it is important to ensure that the electric lighting is turned off when daylight provides adequate illumination. This is achieved by the use of appropriate lighting controls and involves some degree of automation.

The switching circuits for lights have been designed based on a computer simulated lighting grid. Areas with similar light levels are to be located on the same circuit. The lights will be time-programmed to be switched on or off, based on ambient light levels.

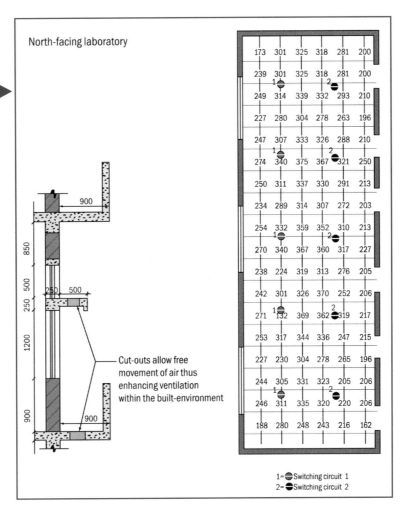

A plan of north-facing laboratory showing artificial lighting levels in a grid of 1 × 1 m at a workplane height of 0.9 m. Artificial lighting has been designed for an average level of 250 lux with twin-tube mirror optic fixtures with 36-W high efficiency tubelights and high frequency ballasts. The combined daylighting levels and artificial lighting levels formed the basis for control strategies for lighting

Efficient lighting equipment

Compact fluorescent lamps (CFLs) are used in place of incandescent lamps. Comparable light output may be obtained from CFL for only 20%–30% of wattage required for incandescent lamps. Maintenance costs are reduced due to long rated life (about eight times more). In the case of fluorescent lamps, 25 mm diameter slim tubelights of 36 W are used in place of 38 mm diameter 40 W lamps, consuming eight per cent less electricity with approximately same output. Optimized use of halogen spotlights for accent lighting and metal halide lamps for external lighting completes the range for an energy-efficient lighting solution.

Renewable energy systems and waste management techniques

Solar photovoltaics

Solar passive systems are built in to the design of the WBPCB building. Computer simulated models have shown a 40% saving in energy consumption over a conventionally designed building of same size and function. Furhter, it is proposed to install a 25-kWp solar PV power plant on the roof. The aim is to fulfil the basic electricity requirement for lighting through the use of solar photovoltaics.

Treatment of waste water

The WBPCB and the forest department buildings, which are located in the same complex, are estimated to reclaim 22 500 litres of water per day. This will be used for flushing toilet cisterns and gardening, as requirement of water in the complex for flushing and gardening is 27 000 litres a day.

Rainwater in the site will not be allowed to run off but will be collected in a water body in the complex. Creation of a water body and sufficient plantations will not only have a cooling effect but also control dust to create a unique and attractive ambience.

Project details

Project description Partially conditioned office building on a busy traffic intersection in Kolkata.
Building/project name Office-cum-laboratory building for West Bengal Pollution Control Board
Climatic zone Warm and humid
Building type Office-cum-laboratory building
Architects Ghosh and Bose & Associates Pvt. Ltd
Energy consultant TERI (Tata Energy Research Institute), New Delhi
Year of start/completion 1996–1999
Client/owner West Bengal Pollution Control Board
Built-up area 4500 m²

Design features

- Optimum orientation of planform
- Solar passive features include optimum window disposition and sizing to allow maximum daylighting, while minimizing adverse thermal effects
- Switching circuits for lights have been designed based on a computer-simulated lighting grid
- Energy-efficient lighting techniques have been adopted
- Shading devices are specifically designed for different wall orientations to control the glare and also reduce the thermal load on the building
- Techniques evolved to treat waste water

Silent Valley, Kalasa

Architect K Jaisim

A unique example of a sustainable habitat, which puts minimum pressure on the environment

Editor's remarks

Away from the madding crowd and close to nature, the Silent Valley resort is a good example of a sustainable habitat with a minimum environmental footprint. Conservation of scarce resources and merger of built environment with nature were the primary objectives of the architect.

Positioned on the eastern slope of the Western Ghats of Karnataka and hugged all round by tall mountains is the Silent Valley, a resort near Kalasa, on the way to Kudremuk. The sprawling resort spread over an area of 1.31 hectare houses cottages, tents, a conference hall, and other ancillary facilities. The built environment is beautifully merged with the natural landscape and responds to the human requirements by use of resource-efficient eco-friendly solutions. Efficient use and reuse of various resources was the primary objective of the architect. The scenario is indeed breath taking and has the effect of taking one's imagination to unbelievable heights.

▲ A view of the cottages with roof made of country tiles and walls of local mud block

Planning and materials of construction

The cottages are partially sunk into the ground to take advantage of the thermal storage capacity of the earth. With the objective of designing a sustainable habitat, the architect has used several innovative techniques. Efforts have been made to retain all the existing mature plants and trees. The buildings follow the existing contour and interplay with the natural landscape. Walls are of solid mud blocks, pillars are of local waste timber, and roof is made of locally available country tiles. Instead of indiscriminately chopping of trees to give way to construction activity, the mature plants have been retained and made a part of the building itself. The roof also allows light to permeate and illuminate the rooms, thereby reducing lighting needs during daytime. Natural tree canopies provide excellent shading for the

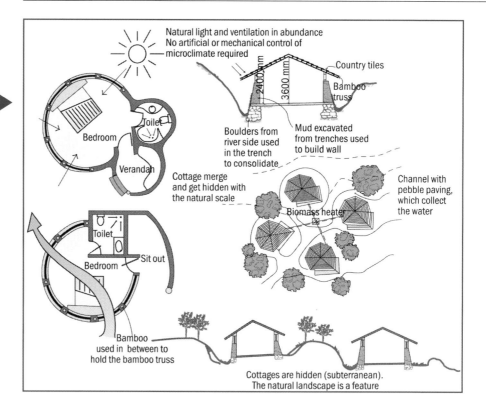

Diagram showing use of on-site sources and sinks for thermal comfort, innovative planning to reduce wastage, and use of local materials

houses. The natural contours have been retained and used effectively to minimize effects of vehicular pollution as well as for protection against strong winds. The restaurant of the resort, which is a multi-deck structure, follows the natural contour and is designed without the conventional wall; the hills are its walls. The roof made of translucent sheets admits daylight round the daytime. The buildings of the resort predominantly uses local materials and labour, which is one of the primary components of sustainable architecture.

The restaurant with translucent roof has hills as its walls

Water conservation

Nature plays an important role in the scheme of things here. Streams that flow through become an important ally of modern technology. Water harvesting is one of the major areas that needs attention in the present-day scenario of scarce resources. The lined channels around the cottages of the Silent Valley guide the rainwater flow, which gets collected in a tank and is used for irrigation.

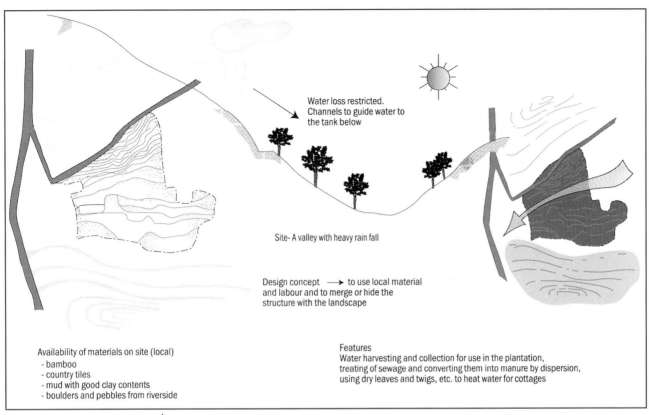

▲ Rainwater harvesting and reuse is a key feature of the resort

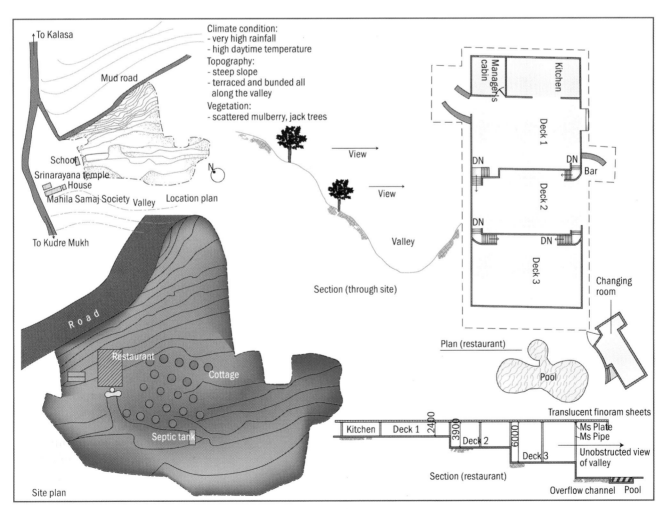

▲ Detailed plan and section of restaurant. Site plan also highlights location of the restaurant

Waste treatment

The disposal of waste is a major national issue, as it proves to be hazardous for everyone's health. However, at the Silent Valley, the waste matter nourishes the tall trees that abound the area. Mulching of leaves provides nutrition for the plants. The waste from the cottages is passed through a septic tank; the outflow is passed through a bed of dispersion channels consisting of eucalyptus trees.

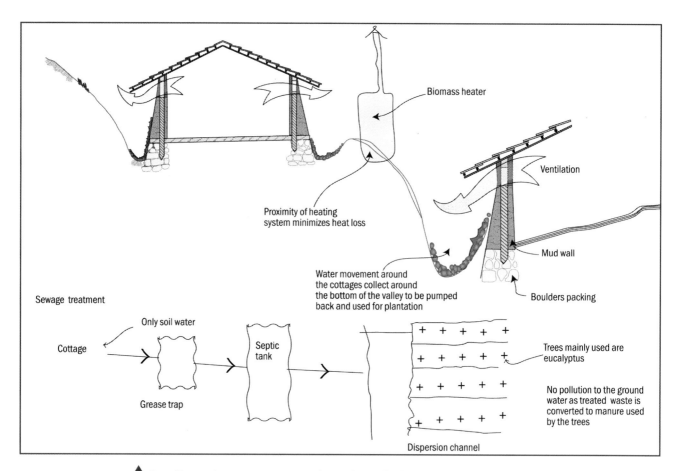

▲ Use of innovative waste treatment and water harvesting leads to maximum resource conservation. Biomass heaters heat water for the cottages

Installed systems

As mentioned earlier, nature has been given prime importance by the architect, thereby minimizing the need for conventional technology. There is no provision for air-conditioning. Here, the benevolence of nature is evident everywhere. Minimum electrical load is required, primarily for reading, security, and communications. Loads are nominal during daytime, as light infiltrates through the dexterously designed roofs. Natural tree canopies help shade the area in a remarkable way.

A centrally located biomass water heater uses dry leaves and twigs to heat water for the cottages. The centralized location of the biomass heater reduces heat loss.

Feedback

'As architects / ecologists, we are thrilled,' chime the architects at Jaisim Fountainhead. 'This is a place that truly houses people from all walks of life—the rich businessman, an average family, children, the adventurous trekker, the philosopher and the artist.'

At a glance

Project details

Building/project name Silent Valley – Resort – Kalasa
Site area 1.31 hectares
Built-up area 1600 m^2
Building type Resort
Architect K Jaisim
Year of start/completion 1997–1999
Cost Building cost is Rs 2800/m^2 (inclusive of civil, electrical, and sanitary cost), and the total project cost is Rs 6000/m^2.

Design features

- Cottages partially sunk into ground
- Use of locally available materials and does away conventional walls, roof, floors, doors, and windows
- Merges with nature instead of trying to modify natural landscape to suit the building needs
- Does away with energy-intensive space-conditioning and lighting systems; makes ample use of daylight
- Waste water used to nourish plants that provide excellent shade
- Biomass heater for water heating
- Water harvesting

Vikas Apartments, Auroville

Editor's remarks

Resource efficiency and community participation are key to energy efficiency. It has been aptly demonstrated in this building, which has used climate-responsive building design and elements, appropriate building technologies, renewable energy technologies, and waste management techniques. Efficiency is being maintained at end-use by conscious use of various resources and systems.

Architect Satprem Maini

Architect Satprem Maini designs residential apartments with a holistic approach for a collective of clients

A habitat is wider than a house and includes the surroundings, the neighbourhood, the infrastructure used for the welfare of all, the environment, etc., and also deals with social relations and patterns. Thus the aspiration of a few friends to live a lifestyle related to the philosophy of Sri Aurobindo and Auroville culminated in this building with 23 residential apartments housing 50 people, and common facilities. Apart from the specific spirituality, the creation of the community aimed to share the financial resources — to build for everyone and not relate the finished product with the financial participation. Constructed in phases, the building makes exclusive use of appropriate building technologies (earth and ferrocement), renewable energy (solar and wind), and ecological water management (water harvesting and waste water treatment).

Vikas: part plan and ancillary facilities. Building oriented so as to catch the summer day and night wind. Oriented longitudinally along east–west axis with shaded openings along north–south for cross-ventilation and reducing summer gains

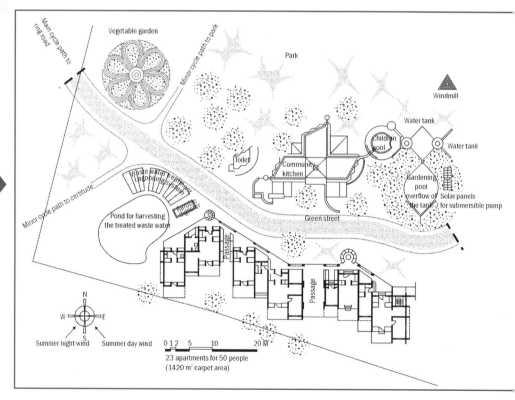

Appropriate architectural design

The building is oriented to catch the summer wind. Natural cross-ventilation is improved by increasing the wind velocity with the help of pier walls oriented at 45 degrees to the predominant wind direction. Solar chimneys are integrated with the building structure to create a natural draft that is specially refreshing at night. The basement floors, being only 1.2 m underground, receive a lot of daylight and at the same time remain cool in the summer.

Appropriate building technologies

Vikas apartments are built with stabilized rammed earth foundations (with five per cent cement) and CEB (compressed earth block) (with five per cent cement) walls, vaults, and domes. Some walls have been built using stabilized rammed earth with five per cent cement. The soil for building has been extracted from the waste water treatment pond and the garden tank, while in the third apartment building with a basement floor, the excavated soil is used for building. Ferrocement roof channels have been used for some floors while doors and shelves are of ferrocement. Concrete, glass, steel, etc. have been sparingly used.

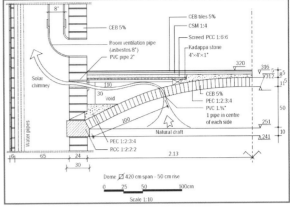

Section showing innovative roof construction for natural draft by a solar chimney

Renewable energy systems

The apartments employ renewable energy sources and conservation techniques. The collective kitchen has a solar water heating system that caters to the cooking needs of the community kitchen. Lighting needs of the apartments are met by solar photovoltaics (installed capacity 2.04 kW). Alternative pumping using two solar pumps and a windmill (for pumping water from depths of 32–35 m) caters to the needs of 100 people who use the water with care.

Wind mill for water pumping and solar photovoltaic for lighting

Water management

The carefully designed landscaping aims at a zero run-off of the monsoon rains. The overflow of the windmill and the solar pumps is used to water the garden. Biological waste water treatment is done with the help of an aerobic process, called lagooning. The sun and aquatic plants do the treatment and after eight days the treated water can be used for gardening (not for drinking).

Till today, the development of Vikas apartments has been exclusively based on appropriate building technologies, renewable energies, and ecological water management. Beyond this alternative material implementation was

Green cover on terrace and west walls reduce heat gain

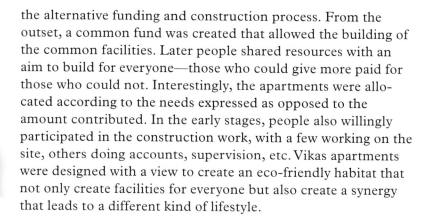

Vikas building: pier walls and solar chimney to augment natural airflow in the building

the alternative funding and construction process. From the outset, a common fund was created that allowed the building of the common facilities. Later people shared resources with an aim to build for everyone—those who could give more paid for those who could not. Interestingly, the apartments were allocated according to the needs expressed as opposed to the amount contributed. In the early stages, people also willingly participated in the construction work, with a few working on the site, others doing accounts, supervision, etc. Vikas apartments were designed with a view to create an eco-friendly habitat that not only create facilities for everyone but also create a synergy that leads to a different kind of lifestyle.

Performance

About 0.6 to 0.8 kWh of renewable energy is consumed per month in the bachelor's accommodation. The residents are satisfied with the thermal comfort and find that summers have rarely been too hot. However, they do find that daylighting is inadequate at times and there is a necessity to augment it.

Installed renewable energy systems and waste management techniques

- Solar photovoltaics for lighting (57 panels of different capacity, totalling 2040 W for 23 apartments)
- The solar water heating system provides hot water only to the kitchen
- Water pumping using a windmill
- Solar water pumping with submersible pumps
- Waste water recycling using lagooning
- Solid waste management with Auroville Eco Service

Maintenance

To maintain this eco-friendly habitat, the architects have issued guidelines to all the residents for the appropriate usage of renewable energy equipment installed in the apartments. A checklist of items that demand periodic checking and maintenance (daily, weekly, monthly, and yearly) has also been given to the residents. It, however, remains to be seen that the residents are not tempted in future to change their style of living in Vikas, which would disturb the harmony and resource sharing, which are the remarkable and essential features of the project.

Design features

- The buildings are oriented longitudinally along the east–west axis with openings along the north–south for cross-ventilation and reducing summer gains. Pier walls oriented at 45 degrees to the predominant wind direction further aid cross ventilation
- The buildings are partly sunken with adequately daylit basement floors (1.2 m deep) that are cool in summer (earth stabilizes internal temperature)
- Soil excavated for construction has been used in making earth-blocks for the buildings thus reducing the embodied energy
- Solar chimneys are integrated with the building structures creating a natural draft that add to the ventilation
- Fenestration with overhangs have been adequately designed to get enough daylight and cut off direct gains
- Terrace gardens and creepers on the west façade reduce cooling loads
- Energy-saving compact fluorescent lamps of 9 W and 6 W have been used for lighting
- Space conditioning is achieved through natural cross-ventilation

Project details

Project description 23 residential apartments housing 50 people, and common facilities
Building type Residential
Climate Warm and humid
Built in area 1420 m²
Owners/clients Collective of clients
Architect Satprem Maini
Period of construction 1992–1999

Pier walls and solar chimney for cross-ventilation

Temperature check for Auroville

(see Appendix IV)

Cold	Cool	Comfortable	Warm	Hot	
		●			January
		●			February
			●		March
			●		April
				●	May
				●	June
			●		July
			●		August
			●		September
			●		October
		●			November
		●			December

Add the checks

		4	6	2

Multiply by 8.33% for % of year

Comfortable	33
Cooling	67

La Cuisine Solaire, Auroville

Architects Suhasini Ayer Guigan and Anita Gaur

A collective solar kitchen demonstrating use of solar energy for community cooking and use of appropriate technologies

Editor's remarks

This collective solar kitchen is an example of reducing strain on conventional energy in buildings by efficient structural design, reducing use of energy-intensive building materials, and use of appropriate technologies for construction. The solar kitchen demonstrates the innovative use of solar thermal energy for cooking meals for 1000 people, thrice a day.

This project was built to demonstrate the use of solar thermal energy in steam generation in cooking meals thrice a day for about 1000 people. The project also demonstrates the appropriateness of compressed earth blocks and ferrocement roofing channels supported by an innovative channel truss beam.

The project will thus support organic farming within Auroville and in local villages by being the main purchaser for their products used for meals prepared in the kitchen. At a later stage, the steam will also be used for food processing and laundry service.

▲ A view of the south-western facade showing solar orientation

Steam is generated using a bowl that is a 60-degree section of a sphere of 18.65-m diameter. The rays of the sun are concentrated on a boiler that is suspended from an arm that is pivoted at the centre of the sphere. The boiler contains a coiled metal pipe that will carry thermic fluid, which will, in a heat transfer chamber, convert water to steam. This boiler moves with the sun, taking into account both the daily movements and the seasonal shift in the position of the sun. However, unlike solar photovoltaic energy, the performance drops drastically in the case of clouds. Therefore, the system is designed as a hybrid system, which means that it is coupled with a conventional boiler in case of temperature drops in the solar bowl.

Solar dish made of prefabricated ferrocement sections assembled in situ and grouted. The resulting structure is then plastered to achieve a smooth finish on which mirrors with the quality of float glass are stuck ▼

The solar bowl is built using prefabricated ferrocement elements shaped as sections of the sphere (like the segments of an orange) that are assembled in situ and grouted. The resulting structure is then plastered to achieve a smooth finish on which mirrors with the quality of float glass are stuck.

Use of appropriate technologies and passive solar concepts

Built on a grid of 2.5 × 2.5 m, this 1700-m² building is built with compressed earth blocks and is a load-bearing structure.

Foundations

The composite foundation technique comprises three layers. Trenches were dug to varying depths depending on the load of the building—between 0.75-m depth for the store rooms and 1.25-m depth for the kitchen. The first layer, with a depth of about 20–25 cm, was composed of sand and pebbles in a dry mix compacted using a hand-held rammer. The second layer is of blue metal of size 40–45 mm mixed with stabilized earth mortar and compacted. The third layer is of granite blocks of 300–350-mm size in random rubble masonry with stabilized earth mortar up to ground level. Above ground level and up to the plinth, blocks are used for construction.

Walls

All walls and pillars are in compressed earth blocks (cement stabilized with five per cent content) manually manufactured using the Auram 3000 block maker. The external wall surfaces are treated with a water repellent paint and internal surfaces are painted with a cement-based paint of the required colour.

Roof

Several roofing systems have been used. The roof, spanning 10 m, over the main kitchen area is of long-span ferrocement channels prefabricated at site and installed manually at 4.5 m above the floor level. The dining hall has doubly-curved shell roofing where the prefabricated ferrocement shells are used as lost shuttering.

The store rooms and auxiliary areas have the normal ferrocement channels up to 6.5 m span. Solar chimneys have been incorporated in the kitchen and dining hall to enhance natural ventilation. The vents provided for solar chimney also provides diffused daylight.

The roof, spanning 10 m, over the main kitchen area, is of long-span ferrocement channels prefabricated at site with solar vents above lintel incorporated to create a natural updraught to ventilate the kitchen and dining hall

Doubly-curved shell roofing for dining room

Openings

Openings are cast in situ RCC (reinforced cement concrete) using ferrocement prefabricated elements as lost shuttering thus saving on finishing plaster. The windows are recessed into the pillars where the lintel acts as a sunshade. Above these lintels, solar vents have been incorporated to create a natural up-draft to ventilate the kitchen and dining hall. All the windows and doors are of steel grills with wire mesh and sliding glass shutters.

Flooring

The flooring in the dining and kitchen areas is in Shabad stone while other areas have coloured cement flooring.

Roof insulation and waterproofing

First, a thermal insulation layer of broken bricks mixed with lime is laid over the roofing channels. This is treated with a fermented solution of jaggery and *Terminalia chebula* (Kaddukai) nuts which is poured over the brick lime jelly and beaten using wooden mallets. Reject tiles in cement screed are then laid as a waterproofing layer.

Staircases

The main spiral staircase is in prefabricated RCC with elements designed to be assembled on site. Staircases within the building for office or roof access are also prefabricated in RCC, inserted during construction.

Overhangs

These are incorporated in the lintel by recessing the openings. The lintels, therefore, have an L profile and the wall moves outside above the lintel level.

Waste water recycling

The soil was excavated at site for making compressed earth blocks for wall construction and the resulting hole used for the waste recycling pond of the root zoning technique. Waste water is recycled using an imhoff tank and baffle reactor with a polishing tank to reuse the water for gardens.

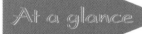

Project details

La Cuisine Solaire demonstrates that a solar concentrator can produce enough steam to cook for 1000 people a day.

Architects Suhasini Ayer Guigan and Anita Gaur
Consultant for the bowl Chamanlal Gupta, Sri Aurobindo Ashram
Contractor / builder Local labour trained and managed by the architect
Covered area 1700 m²
Project duration 1994–1997
Approximate cost Rs 12.5 million (inclusive of cost of the solar concentrator)

Construction techniques for the building

- Composite foundations comprising three layers
- All pillars and walls are of compressed earth blocks stabilized with five per cent cement
- The roof over the main kitchen area is a 10 m long-span ferrocement channel prefabricated on site
- The dining hall has a doubly-curved shell roof where the prefabricated ferrocement shells are used as lost shuttering
- The store rooms and auxiliary areas have the normal ferrocement channels
- Solar chimneys have been incorporated in the kitchen and dining hall to enhance natural ventilation
- Openings are cast in situ RCC using prefabricated ferrocement elements as lost shuttering thus saving on finishing plaster
- Thermal insulation is in broken fired bricks mixed with lime treated with a fermented solution of jaggery and local nuts
- Waterproofing is done with reject tiles in cement screed

Sponsors The Ministry of Non-conventional Energy Sources for the solar bowl and the hybrid system, Housing Urban Development Corporation for the dining hall section, the Foundation for World Education (USA), The Stichting de Zaieer (Holland), and the Auroville Foundation.

Temperature check for Auroville

(see Appendix IV)

Cold	Cool	Comfortable	Warm	Hot	
		●			January
		●			February
			●		March
			●		April
				●	May
				●	June
			●		July
			●		August
			●		September
			●		October
		●			November
		●			December

Add the checks

		4	6	2

Multiply by 8.33% for % of year

Comfortable	33
Cooling	67

Kindergarten School, Auroville

Editor's remarks

Built in a warm humid climate, this kindergarten school is a low-cost, low-energy building. Specially designed roof tiles adopted for daylighting and ventilation are unique to this small school building.

Architect Suhasini Ayar Guigan

Eco-friendly low-cost construction techniques and simple passive solar features for a Kindergarten School in warm, humid climate

The Kindergarten School has been designed for Sri Aurobindo International Institute of Educational Research for approximately 60 children between the age of 3 and 6 years. Funded by the Ministry of Human Resources Development, Government of India, the school has been designed to create an ideal learning environment through play spaces.

 Recessed windows, extended roof for shading walls and windows, ventilated roof by means of special tiles designed for escape of hot air

Passive solar features

The classrooms have load-bearing walls of earth and left with a natural finish. However, the lintels, which is a shading device-cum-beam to support the tile roofs, are painted in different colours thus enabling the children to identify with their group space. The colour of the rooms is in relation to their location in terms of the movement of the sun. The rooms are painted either in cool or warm colours depending on the cardinal direction and how much of the reflected or direct sunlight enters each classroom. For example, the kitchen/dining space is in cold blue as it is not in direct sun when it is used at mid-day. But as it is also the hottest time of the day and the children would be tired after a whole morning, this colour is more refreshing to the senses than a warm colour. There are a number of multipurpose rooms for craft, theatre, and music, besides classrooms, that open onto the gardens that are roofed to provide shade. Creepers and plants provide shade in the transitional spaces. Recessed windows provide sun and rain protection as well as cosy window seats for the children. The roof is made of burnt clay tiles on an understructure of prefabricated reinforced cement concrete beams. Special glass tiles and ventilator tiles in the roof allow hot air escape and at the same time provide an interesting play of light.

Construction techniques

Foundations

The building has rammed earth foundations. As the soil was not of sufficiently good composition, the earth excavated from the trenches was mixed with cement (five per cent) and with sand. The plinth is in CEBs (compressed earth blocks) with a DPC (damp proof course) of 1:3 cement mortar.

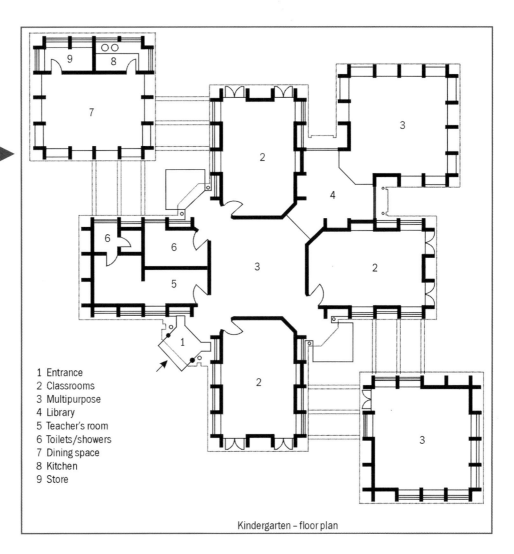

Introvert planning around a central hall, with classrooms on the periphery. The central hall with ventilating tiles in roof creates natural ventilation draft by allowing hot air to escape and forcing in cooler ambient air through the classroom windows

1 Entrance
2 Classrooms
3 Multipurpose
4 Library
5 Teacher's room
6 Toilets/showers
7 Dining space
8 Kitchen
9 Store

Kindergarten – floor plan

Walls

All the pillars and walls are in CEBs, which are cement stabilized with five per cent cement content. These were manufactured manually using the Auram 3000 block maker. The soil used was excavated on site and the resulting hole is used to hold the rainwater from the roofs and the site.

Wall finish

Both the external and internal wall surfaces are left natural without any paint and water repellent finishes.

Roof

The pyramidal roof is of Mangalore tiles with an understructure of prefabricated RCC rafters manufactured on site and assembled in situ with grouting of the joints using a combination of ferrocement and RCC poured *in situ*.

Openings

All the openings of the buildings have an RCC element that is a combination of three things – lintels, overhangs, and gutter – for the channelling of the rainwater from the pyramidal tile roofs.

Doors and windows

All the windows are made with steel grills with glass fixed on them. The doors are a combination of teakwood recycled from demolished houses in Pondicherry and welded mesh. The cupboard doors have transparent glass fixed in wood frames.

Overhangs

The overhangs are the lintels and the ring beam as earlier mentioned. These hold the RCC rafters that have been given a 'U' profile to also act as a gutter.

▲ Special glass tiles for an interesting play of light, and ventilator tiles to allow hot air to escape

Flooring

The internal flooring is cement coloured with red oxide laid on a cement screed of 1:10. The toilets have ceramic tiles both for flooring and as dado.

Waterproofing

The tile roof because of its slope does not require any further waterproofing but the 1-m-deep overhangs-cum-lintels have a waterproofing plaster in cement mortar with waterproofing compound.

External paving

The garden paths and the sit-outs are paved using CEBs.

User feedback The ventilation inside the classrooms as such is good with the airflow being sufficient for comfort. These rooms have ceiling fans installed but they are rarely used, as the school is operating only between 8.00 a.m. and 1.00 p.m. In the late afternoon, it becomes a little uncomfortable. There is no problem with daylighting even in the dark monsoon period as enough window and skylights have been provided. The children have been happy with the building.

Kindergarten School, Auroville

At a glance

Project details

Project Kindergarten School, Auroville
Architect Suhasini Ayar Guigan
Plinth area 560 m²
Climate Warm and humid
Contractor Local labour trained and managed by building centre, Auroville
Project duration 1991–1992
Cost Rs 1.5 million (approximately)

Design features

- Eco-friendly energy-efficient construction techniques:
  - Use of rammed earth foundations
  - Pillars and walls using five per cent cement stabilized compressed earth blocks
  - Mangalore claytile roof with an understructure of prefabricated RCC rafters
  - Doors of teakwood recycled from demolished houses in Pondicherry and welded mesh
  - External paving using compressed earth blocks.
- Specially designed roof tiles for daylighting and ventilation
- The excavation pit left after excavating soil for making of compressed earth blocks was used for rainwater harvesting
- Colour schemes as per room location with respect to solar path.

Temperature check for Auroville

(see Appendix IV)

Cold	Cool	Comfortable	Warm	Hot	
		●			January
		●			February
			●		March
			●		April
				●	May
				●	June
			●		July
			●		August
			●		September
			●		October
		●			November
		●			December

Add the checks

		4	6	2

Multiply by 8.33% for % of year

Comfortable	33
Cooling	67

Visitors' Centre, Auroville

Architects Suhasini Ayer Guigan and Serge Maini

This building demonstrates use of appropriate building technologies and methods of construction, land reclamation and afforestation, renewable energy technologies, and water and waste management techniques.

Editor's remarks

The Visitors' Centre at Auroville has been built to showcase possibilities of energy conservation by use of various appropriate technologies in construction. Special emphasis has been laid on daylighting and ventilation techniques to reduce energy consumption at end use.

The Auroville Building Centre (AV-BC) is a research-cum-training organization in appropriate building technology that forms part of the national network of building centres set up with HUDCO's (Housing and Urban Development Corporation's) assistance. The AV-BC specializes in earth and ferrocement technologies and also produces a whole range of prefabricated ferrocement elements. The Auroville Visitors' Centre has been designed and built by the AV-BC as a demonstration project for alternative technologies.

The entrance arcade of the Visitors' Centre is spanned by arches in compressed earth blocks

The project is located in the southern Indian township of Auroville and is the reception centre for hundreds of visitors who visit Auroville everyday. It is a demonstration complex for alternative technologies such as appropriate building technologies, land reclamation and afforestation, renewable energy technologies, water management, and waste recycling techniques.

This building acted as a training site for the local villagers who built it, thus learning soil block construction, ferrocement roofing techniques, earth construction techniques such as arches, domes, and building techniques.

Aims

Various types of people use this building. Thus it had to be inviting for visitors who would be drawn to explore the space and become aware of the possibilities offered by various appropriate technologies. The building has an information office, exhibition spaces, restaurant, video-room, shop for handicrafts, conference facilities, toilets, and an open-air amphitheatre, the last serving as a focal point.

In order to reduce the cost and simplify the construction, a grid pattern of 4 × 4 m was adopted in which the pillars are load-bearing and all openings are arched or corbelled.

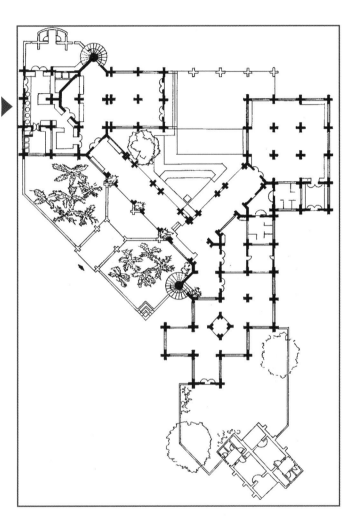

A shallow modular grid has been adopted, which aids uniform daylighting, reduces structural costs and the construction process also has thereby been simplified

Materials of construction

The locally available fired bricks are of poor quality but the soil is suitable for making blocks. The built-up area is about 1200 m². Soil stabilized blocks are used to build the pillars and arches. Some spaces are covered with domes constructed with soil stabilized blocks (24-cm thick), which at times has a first floor above it. Other spaces are covered with prefabricated ferrocement channels that have a first floor above it at times. Timber has not been used in the building because of its high cost, vulnerability to termites, and contribution to reducing national forest cover.

Daylighting and ventilation

Special emphasis is laid on natural lighting and ventilation as the building taps only renewable sources of energy and is not connected to the grid of the Tamil Nadu Electricity Board.

The climate is warm and humid. Therefore, natural ventilation is necessary for comfort. However, during the hottest part of the year when there is almost no breeze, an integrated solar chimney causes wind draft in the room below. This is a demonstration passive solar ventilation technique (refer to chapter 'Energy efficiency in architecture: an overview of design concepts and architectural interventions' pp. 1–18). Built using low-cost techniques to reduce the cost per square metre, the building creates a viable alternative to conventional construction techniques.

The energy-efficiency in this project is governed by the choice of materials and construction techniques.

Space planning and construction techniques

Spatial planning

The spaces have been planned in a modular grid of 4 × 4 m and are of shallow depth to aid uniform daylighting. The modular grid of 4 × 4 m has reduced structural costs significantly and the construction process has been simplified. The pillars are load-bearing and in CEBs (compressed earth blocks). All openings are arched or corbelled thereby reducing structural costs.

Foundations

Composite foundations comprising three layers are used due to the clayey soil conditions. Trenches were dug up to the same soil type everywhere (which meant that they were of unequal depth). The first layer of about 20–25-cm depth was composed of sand and pebbles in a dry mix, compacted using a hand-held rammer. The second layer is blue metal of 40–45-mm size mixed with stabilized earth mortar and compacted. The third layer comprises granite blocks of 300–350-mm size laid in random rubble masonry in stabilized earth mortar up to ground level. The plinth is in dressed granite blocks with a damp proof course in bitumen.

Walls

All pillars, arches, and walls are in CEBs, which are cement stabilized with four or five per cent cement content as required. These were manufactured manually using the Astram maker. The soil used was excavated on site and the resulting hole used for the waste-recycling pond.

The building demonstrates energy efficiency by choice of low energy materials and construction techniques. Use of arches to span openings is a predominant feature of the building

Wall finish

The external wall surfaces are left without any paint or protection. The internal surfaces are given a rendering with cement-based paint of the required colour.

Roof

There are two types of roofs. The domes over all the rooms are made with CEB tiles (24 cm thick) manufactured with the Auram block maker. These domes are built without using any centring and they are a section of a sphere. The corridors and the other such spaces are roofed over with prefabricated ferrocement channels, which were manufactured at AV-BC.

Openings

The openings are either arches or corbels in CEBs. The arches were built using a centring in steel. The centring was removed immediately on completion of each arch. The PCC corbel stone used at the base of the arch was needed only because the design of the arch was too flat (for aesthetic reasons).

Doors and windows

All the windows and some doors are made of steel grillwork with either wire mesh or glass fixed on them. Doors that are exposed to wet areas are made of ferrocement.

Staircases

The two spiral staircases are of prefabricated reinforced cement concrete designed to be assembled on site.

Overhangs

Overhangs are of prefabricated ferrocement installed on brackets built into the wall with scaffolding poles at the time of construction so that the overhangs do not crack with the settlement in the building.

Outdoor paving

The paving in the entrance of the building is a combination of waste terracotta roofing tiles, Cuddapah stones, and granite slabs. The paving in the courtyard-cum-amphitheatre is done by CEB tiles, similar in thickness to the ones used in the dome construction, with the same stabilization as the wall tiles.

Flooring

Internal flooring is done by unpolished Cuddapah stone tiles of green colour while toilets and other wet areas have ceramic tiles.

Waterproofing

The waterproofing over the domes is done on an insulation layer made of broken fired bricks mixed with lime. This is treated with a fermented solution of jaggery and *Terminalia chebula* (*Kaddukai*) nuts which is poured over the brick lime jelly and beaten in using wooden mallets. On this, a plaster of shell lime is done and a final rendering of lime wash is given. Flat roofs on ferrocement channels have an insulation of brick jelly lime, laid to a minimum thickness of 3 inches, over which there is a waterproofing plaster of lime-cement finished with crazy mosaic of broken ceramic tiles.

At a glance

Project details

Site location Auroville

Climate Warm and humid

Covered area (plinth) The approximate covered area of the project is 1200 m²

Sponsors The main sponsors of the project were Housing and Urban Development Corporation, Foundation for World Education, Stichting de Zaieer (Holland), Indian Navy, United Nations Centre for Human Settlements (Nairobi), Ministry of Non-conventional Energy Sources, and Auroville Foundation

Consultants Centre for Scientific Research (Renewable Energies)/Aureka/Ray Meeker/Craterre, France

Contractor/builder Local labour were trained and managed on site

Project duration 1989–1992

Approximate cost Rs 3.5 million

Design features

- Spatial planning has been done in a modular grid of 4 × 4 m, reducing structural costs and aiding daylighting
- Use of alternative and innovative construction techniques has reduced embodied energy immensely. Some of these are stone foundation, walls in compressed earth blocks, roof in compressed earth tiles, and ferrocement channels, prefabricated RCC staircase, ferrocement external doors, etc.
- Use of solar chimney for ventilation, suited to the warm and humid conditions of Auroville
- Excavation pit left after excavating soil for making of compressed earth blocks and tiles used for waste water recycling pond

Temperature check for Auroville

(see Appendix IV)

Cold	Cool	Comfortable	Warm	Hot	
		●			January
		●			February
			●		March
			●		April
				●	May
				●	June
			●		July
			●		August
			●		September
			●		October
		●			November
		●			December

Add the checks

		4	6	2

Multiply by 8.33% for % of year

Comfortable	33
Cooling	67

Computer Maintenance Corporation House, Mumbai

Architects Vinod Gupta and Rasik Bahl

An energy-efficient, intelligent, hi-tech building in the busy metropolis of Mumbai

Editor's remarks

Located in a tight urban setting, the built form of the CMC House was predetermined by strict zoning regulations. However, the architect succeeded in designing an energy-efficient and intelligent building by thoughtful integration of architecture with interior design and services.

The CMC (Computer Maintenance Corporation) building at Bandra, Mumbai, is an intelligent, hi-tech, and energy-efficient building. Architecture, interior design, and services are integrated to achieve these objectives.

The basic requirements of the building emerged from a series of meetings between the designers and the 'users committee' of the CMC, regarding office space organization, the designs of the library, meeting rooms for interaction with visitors, a reception area, museum-cum-art gallery, cafe and recreational facilities for the staff, besides parking and services, which are located in the basements.

To encourage informal interaction among the staff, who would be seated on different floors in the building, it was decided to create small work modules of 10 to 25 persons, each arranged around an atrium. These are arranged in the form of helix, resulting in a building in which floor divisions have disappeared and communication is enhanced, without losing privacy and spaciousness. It even encourages people to take the easy-going stairs instead of the elevator.

All the public spaces such as the reception and meeting rooms, art gallery, computer room, and demonstration room are located at the lower levels, leaving 'six floors' of continuous office space topped by the staff facilities. Car parking and various services are provided in the basement.

The form of the CMC building was determined by the prevailing zoning regulations, which called for a building of 22 × 25 × 30 m dimension.

◀ A view of the CMC building, Mumbai

Design details

The architectural concept divided each floor into four levels connected by short flights of stairs. Three of these levels are office spaces while the fourth is the service module with the elevators and the main staircase. The main circulation route is placed around the atrium in the form of a helix, which creates an interesting spiralling space.

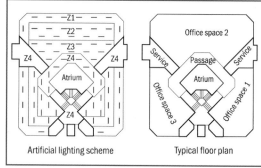

▲ Schematic plans showing central atrium and arrangement of office spaces around them

Building systems

Lighting system

The thoughtfully designed lighting system provides a uniform general illumination in the range 175–200 Lux in the entire working area using low wattage and high efficiency sources such as CFL or fluorescent lamps (40 W) with high frequency electronic ballasts, mirror optic reflectors, and parabolic louvres. Since the work surfaces require higher illumination (~300 Lux), task lighting has been provided with the help of 9 W CFL PL task lamps. Together, these require no more than 10 W of power per square metre of office space. The on/off and dimming of artificial general lighting is controlled by a microprocessor that responds to the available daylight. The lighting of computer areas required special attention in order to avoid glare on the computer screens. Special fixtures with metallic parabolic mirror louvres solved this problem without sacrificing the efficiency of the luminaries. A more uniform level of lighting has been achieved by using closely spaced single-tube light fixtures instead of the common twin-tube fixtures.

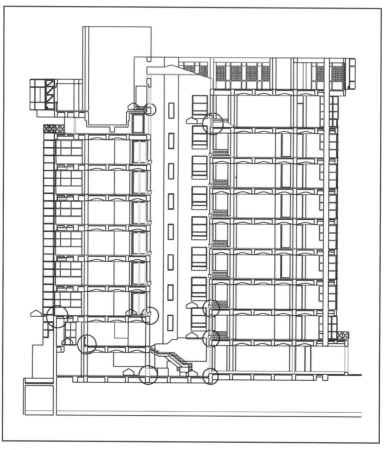

▲ Section through the CMC House

Air-conditioning system

Four independently controlled unitary AHUs (air-handling units) are located at each floor allowing flexibility in use of the building. When the building is not fully occupied (over a weekend for example), it is possible to cool only those sections which are in use.

The HVAC (heating, ventilation, and air-conditioning) plant is located in the basement and mix-matched to allow variable capacity with high efficiency. The entire operation of the plant and the AHUs is controlled by a microprocessor. Apart from providing efficient control over the operations, the microprocessor control system also ensures automatic load management when the building is to be run on a stand-by power generator and full plant operation is not possible. In the event of a fire, unitary AHUs will be switched off, ventilation fans will be started, and lift lobby will be pressurized.

Interior design

In addition to the functional and aesthetic requirements, the materials used in the interior design were carefully chosen with environmental conservation in mind; the tradition of the CMC of using handcrafted natural materials in preference to synthetic materials has been followed.

An open plan landscape office concept has been used all through. In order to reduce noise in the open office, sound absorbent ceiling, partitions, and flooring have been used. The furniture itself is made out of the waste wood from rubber trees, processed to produce a durable and aesthetically pleasing material. The modular sound absorbent dividing screens incorporate a suspension system that allows various modules like table tops, side racks, and storage units to be attached to them. Electricity, telephones, and computer communication cables are integrated with the screens, flooring, wall panels, false ceiling, and also with the furniture. Specially produced coir mat with latex backing has been used to provide a durable and inexpensive floor covering in office.

Special features

Integration of daylighting with the lighting system

Windows located on the periphery are unable to transmit daylight deep into the working areas (11–13 m in the case of this building). A central atrium with circulation routes around the atrium introduces daylight at all floors and effectively reduces the depth of working areas to 9 m.

The daylit atrium at the CMC, Mumbai

The double-glazed peripheral windows are split in to two parts—the upper half is used for daylighting and the lower half for view. The upper window incorporates small light shelves in the form of reflective (mirror coated) venetian blinds operated by microprocessor-controlled motors which tilt the slates at predetermined angles in response to the angle of incidence of sunlight on different faces of the building. The motorized louvres are designed to be automatically adjusted, to reflect sunlight on to the ceiling, which has white painted flat and angled panels.

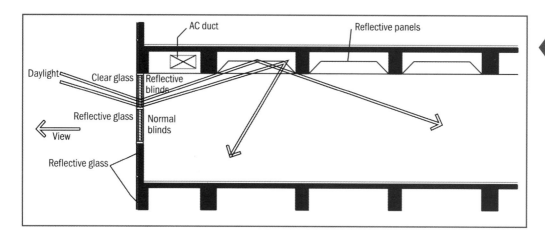

Section of the office space with detailing for uniform daylighting in the office spaces

This method provides more light deep inside the office space and improves daylight distribution near the windows. The lower half of the window is fitted with reflective glass and ordinary venetian blinds, which people can adjust according to their needs. For better maintenance, the blinds are enclosed between the outer and inner sheets of glass. The net effect of these measures is that during daylight hours, practically no artificial lighting is needed.

Insulation for reduction in heat loads

To reduce the cooling load, thermal insulation was provided in all exposed wall and roof surfaces and double-glazed. The windows are sealed to reduce infiltration losses. To allow more efficient plant operation, the air-conditioning load was determined by the more commonly occurring ambient conditions in preference to the normal practice of using the worst combination of ambient factors which might occur on one or two days in the year. The total air-conditioning plant capacity thus worked out to about 17 m² of cooled space per TR (tonnes of refrigeration), which is much better than the Mumbai norm (~10 m² per TR)

Intelligent (hi-tech) building

Integration of computerized management and control in the building design is not just to manage normal building functions but also to achieve a higher level of efficiency. The system controls air-conditioning, lighting, security, and automatic maintenance of equipment and communications. The computers also make the building respond to the outdoor weather conditions to optimize energy use in the building. Computers are also used here to allocate seats to field workers on a day-to-day basis, and for informing the EPABX of the location of the worker. This enables the company to optimize the seating capacity of the building and other facilities.

Energy conservation in the CMC House

The designers provided the first estimates that the CMC building consumes about 25% less energy than the traditionally designed offices in Mumbai. Since the controls for lighting, venetian blinds, and the air-conditioning plant were part of the Building Automation System developed in-house by the CMC itself, it is difficult to give authentic estimates for the additional cost incurred in designing this system and other energy-saving features. The cost of the automation system was 6.5 million rupees and the savings in reduced installed capacity of the HVAC plant were about 3 million rupees.

The microprocessor for controlling the movement of the louvres was installed in some parts of the building only and in most of the building the blinds work no different than in any other conventional building. These blinds are now controlled by a manually operated electrical switch.

Computer Maintenance Corporation House, Mumbai

At a glance

Project details

Project name CMC House, Mumbai
Building type Commercial office
Climatic zone Warm and humid
Client/owner CMC Limited
Architects Vinod Gupta and Rasik Bahl

Consultants
 Project management Stup Consultants
 Structural S V Damle
 Electrical Electrical Consulting Engineers
 HVAC Gupta Consultants
 Plumbing S G Deolalikar
 Mechanical G P Khungar
 Lighting Ashok Gadgil
 Automation CMC Limited & Space Design Associates
 Interiors Carl Christiansson & Space Design Associates
 Energy Vinod Gupta

Start/completion 1989–1992
Total area 5800 m²
Total cost 80 million rupees
Cost of automation system 6.5 million rupees

Design features

- Energy-efficient lighting
- Daylight integration with artificial lighting
- Use of eco-friendly materials for interior design
- Double-glazed windows and wall insulation to reduce building cooling loads
- Intelligent control of air-conditioning, lighting, security, and EPABX system

Temperature check for Mumbai

(see Appendix IV)

Cold	Cool	Comfortable	Warm	Hot	
		●			January
		●			February
			●		March
			●		April
			●		May
			●		June
			●		July
			●		August
			●		September
			●		October
			●		November
		●			December

Add the checks

		3	9	

Multiply by 8.33% for % of year

Comfortable	25
Cooling	75

Appendices

Commonly used software packages in energy-efficient building analysis and design

List of institutions / organizations, architects, and scientists working on energy-efficient buildings

Financial incentives by the Ministry of Non-conventional Energy Sources for the promotion of solar passive architecture

Explanation of the temperature check charts provided in the book

Energy-efficient glazings

Product and services offered by various companies

Appendix I

Commonly used software packages in energy-efficient building analysis and design

Software	Applications
AAMASKY	Skylights, daylighting, commercial buildings
ADELINE	Daylighting, lighting, whole-building simulation, commercial buildings
AkWarm	Home energy rating systems, home energy, residential modelling, weatherization
Analysis Platform	Heating, cooling, and SWH (solar water heating) equipment, commercial buildings
APACHE	Thermal design, thermal analysis, energy simulation, dynamic simulation, system simulation
APACHE-HVAC	Buildings, HVAC (heating, ventilation, and air-conditioning), simulation, energy performance
Awn Shade	Solar shading, awnings, overhangs, side fins, windows
BEES	Environmental performance, green buildings, life-cycle assessment, life-cycle costing, sustainable development
BSim2000	Building simulation, energy, daylight, thermal analysis, indoor climate
Builder Guide	Design, residential buildings
Building Design Advisor	Design, daylighting, energy performance, prototypes, case studies, commercial buildings
Building Energy Modeling and Simulation – Self-learning Modules	Energy simulation, buildings, courseware, self-learning, modelling, simulation
CATALOGUE	Windows, fenestration, product information, thermal characteristics
CBIP Comply	Whole-building performance, building incentives
DOE-2	Energy performance, design, retrofit, research, residential and commercial buildings
EASY: Whole House Energy Audit	Energy audit, residential buildings, retrofit, economic evaluation, DSM (demand-side management)
EcoAdvisor	Online interactive training, online multimedia training, sustainable commercial buildings, lighting, HVAC (heating, ventilation, and air-conditioning)
Energy Profiler Online	Online, energy usage, load profiles, bill estimation
Energy Scheming	Design, residential buildings, commercial buildings, energy efficiency, load calculations
Energy-10	Design, residential, and small commercial buildings
Envest	Sustainable design, green buildings, life-cycle analysis, environmental impact analysis
ESP-r	Energy simulation, environmental performance, commercial buildings, residential buildings, visualization, complex buildings and systems
EZDOE	Energy performance, design, retrofit, research, residential and commercial buildings
FLEX	Lighting, retrofit, commercial buildings
FLOVENT	Airflow, heat transfer, simulation, HVAC (heating, ventilation, and air-conditioning), ventilation
FLUCS	Illumination, daylighting
FRESA	Renewable energy, retrofit opportunities
Green Building Advisor	Sustainable design, green building, energy, water, environment, health, indoor air quality, efficiency
HBLC	Heating and cooling loads, heat balance, energy performance, design, retrofit, residential and commercial buildings
Home Energy Saver	Internet-based energy simulation, residential buildings

Continued...

Software	Applications
HOT2000	Energy performance, design, residential buildings, energy simulation, passive solar
HOUSE	Energy simulation, houses, residences, space-conditioning
ISPE	Solar architecture, passive solar, residential buildings, primer, introduction, educational, reference
LESO-COMFORT	Thermal comfort, load calculation, energy
LESOCOOL	Airflow, passive cooling, energy simulation, mechanical ventilation
LESODIAL	Daylighting, early design stage, user-friendliness
LESOKAI	Thermal transmission, water vapour, building envelope
LESOSAI	Heating energy, energy simulation load calculation, standards
LESO-SHADE	Shading factors, solar shading, building geometry
LifeCycle	Life-cycle cost, economics
Lighting Boy	Lighting retrofit, audit, lighting design, existing buildings
Load Express	Design, light commercial buildings, heating and cooling loads, HVAC (heating, ventilation, and air-conditioning)
LoadCalc Plus	Load calculation, energy cost analysis
NewQUICK	Passive simulation, load calculations, natural ventilation, evaporative cooling, energy analysis
Opaque	Wall thermal transmission, U-value
Overhang Design	Solar, window, overhang, shading
PC-Solar 2.0	Solar shading, external shading, internal shading, solar incidence
Pvcad	Photovoltaic, facade, yield, electrical
PV-DesignPro	Photovoltaic design, tracking systems, solar, electrical design
Radiance	Lighting, daylighting, rendering
RESFEN	Fenestration, energy performance
ShadowFX	Shading calculations, sun modelling, solar shading
Solacalc	Passive solar, house design, building design, building services, design tools
SOLAR-2	Windows, shading fins, overhangs, daylight
SOLAR-5	Design, residential and small commercial buildings
SolArch	Thermal performance calculation, solar architecture, residential buildings, design checklists
SolarPro 2.0	Solar water heating, thermal processes, alternative energy, simulation
SolDesigner	Design, solar thermal, solar hot water, solar heating plants, solar design
Sombrero 2.0	Solar shading, solar radiation, building geometry solar systems
SUN_CHART Solar Design Software	Sunchart, solar position, sun path, shading
SunAngle	Solar, sun, angle
Suncast	Solar shading, insolation
Sunspec	Solar radiation, illuminance, irradiance, luminous efficacies, solar position
SuperLite	Daylighting, lighting, residential and commercial buildings
TownScope II	Solar energy, urban design, visual reasoning
TRNSYS	Design, retrofit, research, energy performance, complex systems, commercial buildings
tsbi3	Energy performance, design, retrofit, research, residential and commercial buildings, indoor climate
VisualDOE	Energy performance, design, retrofit, research, residential and commercial buildings
WATERGY	Water conservation opportunities, energy savings
Window	Fenestration, thermal performance, solar optical characteristics, windows, glazing
Window Heat Gain	Solar, window, energy

Note

This list has been compiled from the web site of EREN (Energy Efficiency and Renewable Energy Network) of the US Department of Energy. The complete tools directory (BTS) describes more than 200 energy-related software tools for buildings, with an emphasis on using renewable energy and achieving energy efficiency and sustainability in buildings. More information on these software may be obtained using the link: http://www.eren.doe.gov/buildings/tools_directory/

Appendix II

List of institutions / organizations, architects, and scientists working on energy-efficient buildings

Institutes and organizations

Anna University of Technology (AUT)
College of Engineering, Guindy,
Chennai – 600 025

Building Technology Division
Department of Civil Engineering
Indian Institute of Technology
Chennai – 600 036

Central Building Research Institute (CBRI)
Roorkee – 247 672

Centre for the Application of Science and Technology to Rural Areas (ASTRA)
India Institute of Science
Malleshwaram
Bangalore – 560 012

Centre of Energy Studies (CES)
Indian Institute of Technology
Hauz Khas, New Delhi – 110 016

Development Alternatives (DA)
B-32, Institutional Area
New Mehrauli Road
New Delhi – 110 016

Energy Systems Engineering
IIT Bombay, Mumbai – 400 076

Gherzi Eastern Limited
16, Mahanirban Road
Kolkata – 700 029

Ghosh Bose and Associates (P) Ltd
8 Harrington Mansion
8 Ho Chi Minh Sarani
Kolkata – 700 071

Housing and Settlements Management Institute (HSMI)
HUDCO House, Lodhi Road
New Delhi – 110 003

Housing and Urban Development Corporation Limited (HUDCO)
India Habitat Centre
Lodhi Road, New Delhi – 110 003

Renewable Energy and Energy Conservation Group
Devi Ahilya University, Indore

Solar Energy Centre
Ministry of Non-conventional Energy Sources
Gual Pahari, Gurgaon District
Haryana

Tata Energy Research Institute (TERI)
Darbari Seth Block, Habitat Place
Lodhi Road, New Delhi – 110 003

University of Roorkee
Department of Architecture and Planning, Roorkee – 247 667

Himachal Pradesh State Council for Science, Technology and Environment
B-34 SDA complex, Kasumpti
Shimla – 171 009

Architects and scientists

Laurie Baker
The Hamlet
Nalanchira PO
Thiruvananthapuram – 695 018

N K Bansal
Centre of Energy Studies (CES)
Indian Institute of Technology
Hauz Khas, New Delhi – 110 016

Gerard de Cunha
Architecture Autonomous
House No. 674 (Opp. Nisha's Play School), Alto Porvorim
Bardez, Goa – 403 521

Manmohan Dayal
B-3/3552 Vasant Kunj
New Delhi – 110 030

Balakrishna Doshi
Vastu-Shilpa Foundation for Studies & Research in Environmental Design
Sangath Thaltej Road
Ahmedabad – 380 054

Chaman Lal Gupta
Solar Agni International
Sri Aurobindo Ashram
Pondicherry – 685 001

Vinod Gupta
Space Design Consultants
K-38 Jungpura Extension
New Delhi – 110 014

Suhasini Ayer Guigan
Auroville Building Centre
Auroshilpam, Auroville – 605 101

Jaisim Fountainhead
175/1 Pavilion Road, 1 Block East
Jayanagar, Bangalore – 560 011

Sen Kapadia
104 Oyster Shell
Juhu Beach, Mumbai – 400 049

Arvind Krishan
Center for Architecture Systems Alternatives
B 4–103 Safdarjung Enclave
New Delhi

Shubhendu Kaushik
C-404 Som Vihar Apartments
Ramakrishna Puram
New Delhi – 110 022

Ashok B Lall
B 25 Chirag Enclave
New Delhi – 110 017
or
2B Ramkishore Road
Civil Lines
Delhi – 110 054

Satprem Maini
Auroville Building Centre
Auroshilpam
Auroville – 605 101

Nisha Mathew and Soumitro Ghosh
Mathew & Ghosh Architects
2 Temple Trees Row
Viveknagar PO
Bangalore – 560 047

Sanjay Mohe
M/s Chandavarkar and Thacker Architects (Pvt.) Ltd
7, Palace Cross Road
Bangalore – 560 020

Nimish Patel
Abhikram
15 Laxmi Nivas Society, Paldi
Ahmedabad – 380 007

Pravin Patel
A-13 Aditya Complex
Opp. Television Station
Off. Drive in Cinema Road
Ahmedabad – 380 054

Sanjay Prakash
Sanjay Prakash and Associates
F- 25 Basement
Lajpat Nagar III
New Delhi – 110 024

Anant Mann and Siddhartha Wig
The Elements
279 Sector 6
Panchkula – 134 101

Appendix III

Financial incentives by the Ministry of Non-conventional Energy Sources for the promotion of solar passive architecture

Solar Buildings Programme

The following are the details of the Solar Buildings Programme, 1998-99 being implemented by Ministry:

Objective

The objective of the Solar Buildings Programme is to promote energy efficient building designs by optimum use of available solar energy (and other forms of ambient energy) in building energy management.

Programme components

Research and development

R&D projects are sponsored to universities, research organizations and other institutions with the objective of developing suitable design techniques and concepts, software packages, materials, architectural instruments, thumb rules etc. for solar efficient buildings.

Training and education

Workshops and seminars are being organized throughout the country for creating awareness, generating public interest and providing inputs about the technology to engineers, academicians, scientists, planners, builders, students, consultants, housing financing organizations and potential house owners. Orientation courses are being organized for architects and builders to make them familiar with the new developments and to motivate them for adopting solar efficient building design concept.

Financial provision for the training and education programme

National/model workshop/ Orientation course	Rs 100 000 each
Seminar/workshop of one day duration	Rs 30 000 each
Refresher/Orientation course of 3 days duration	Rs 60 000 each
Workshop/Orientation course of 2 days duration or a session in international conference	Rs 50 000 each

Awareness programme

This programme envisages creating awareness about the technology through publication of popular literature, books for architects and designers, general publicity through various media etc.

Demonstration programme

To demonstrate the concept of solar buildings, the Ministry accepts proposals for solar building projects for construction from government and semi government organizations generally through State Nodal Agencies. To encourage these organizations for constructing their new buildings on the basis of solar design principles, the Ministry provides the following partial financial assistance:

a) *Preparation of detailed project reports (DPRs)*
   50% of the cost of DPR of a building designed with the help of solar building design principles or 1.5% of the estimated cost of the building with a maximum of Rs 50 000 for each project.

b) *Construction of Solar Buildings*
   Limited to 10% of the cost of the building with a maximum of Rs 10 00 000 for each project.

Other activities

In addition to the above mentioned programmes, activities relating to setting up of data base, review of national building code, development of suitable curriculum for students of architecture etc. are being taken up.

Incentives suggested for private building owners*

Three types of financial incentives are suggested for increasing the rate at which energy conserving measures are introduced in commercial and residential buildings. These are tax-reducing incentives, low-interest loans, and rebate programmes. Tax incentives include reduction of income taxes for business owners investing in energy conservation. It is important to consider incentives for both new building construction and for retrofitting existing buildings.

Another suggestion is to offer accelerated depreciation for energy-efficient lighting, air-conditioning, and water heating equipment.

It is suggested that at least 10 demonstration projects be financed every year with low-interest loans. The financial incentive is necessary to increase the rate at which solar and new technologies penetrate the commercial sector.

In some countries rebates are offered to consumers on purchase of energy-efficient appliances. A rebate is offered for appliances with an energy-efficiency exceeding a minimum value. This will lower electricity demand in

---

*This portion comprises suggestions and recommendations emanated from various workshops, seminars, etc. It does not form a part of the Solar Buildings Programme of the Ministry of Non-conventional Energy Sources, Government of India.

the residential sector by fostering purchases of energy-efficient appliances.

A fourth method of increasing the penetration of energy conservation measures in existing commercial buildings is to encourage alternative methods of financing conservation. In other nations, energy service companies have been established in the private sector. These companies provide energy audits, retrofit capabilities and financing all-in-one package. In order to facilitate the start-up of such companies, the MNES can sponsor several seminars to provide information on the experience of other nations with energy service companies. Invitee to such seminars will include engineering and architectural firms, the financial community, and utility personnel.

Energy service companies audit energy use in buildings to identify the optimal mix of measures for energy efficiency, then install and maintain these measures, often paying for the whole project at no upfront cost to the building owner who is required to pay no more for energy than would have been the case without the efficiency measures. This type of business arrangement is called performance contracting.

There are two kinds of performance contracts: shared saving and energy services contracts. In each case, the actual energy use of the building is compared to what it would have been without the energy-conserving measures. Both parties to the contract have to agree on the methods of comparison including weather corrections, when necessary. In shared savings contracts, the rupee value of the measured savings is divided by some contractually agreed-upon formula between the customer and the energy services firm. If there are no savings, the customer pays his energy bill and owes the vendor nothing. In an energy services contract, the customer pays a fixed amount each month to the energy services company. If the actual energy costs are less than that amount, the vendor keeps the difference. If the actual bill is greater than the flat fee, the vendor must pay the difference.

The earnings of energy services companies depend upon actual energy saved rather than on investment in conservation. This represents a significant advantage compared to using financial incentives based on expenditures on conservation. Additionally, investment funds can be provided by sources in the private sector.

Appendix IV

Explanation of the temperature check charts provided in the book

Evaluation of given climatic conditions at a site is necessary for planning a 'climatically sensitive' layout, giving maximum benefit from good weather and optimum protection from adverse outdoor conditions. Checking each month's temperature relative to human comfort requirements is the first step (Figure 1).

Inference from Figure 1

How severe are your heating or cooling needs?
- 92% of the time in a year you need heating
- 8% of the time of the year you have comfortable conditions
- You do not need cooling at any time of the year

What is 'cold'? Temperature levels below 17 °C down to 5 °C require permanent mechanical heating and benefit from the use of passive solar gains during mild winter conditions.

What is 'comfortable'? When temperature range between 17 °C and 26 °C, the building should quickly react to pleasant outdoor conditions for indoor thermal comfort. Building benefits include energy savings and improved comfort and utilization of spaces around the building for outdoor activities.

What is 'warm'? Temperature periods between 26 °C and 32 °C could mostly be managed by natural cooling techniques when summer conditions are not accompanied by high humidity levels.

What is 'hot'? Temperatures above 32 °C for long periods require cooling. Depending on humidity level, sun, and wind conditions, mechanical air-conditioning will often be the only way to achieve thermal comfort. The building should be protected from the sun and hot air to keep cooling requirements low.

Reference

International Energy Agency. 1989
Passive and hybrid solar low energy buildings: design context
Solar Heating and Cooling Programme (Booklet No. 2)
Paris: International Energy Agency

Figure 1 Temperature check for Leh

Cold	Cool	Comfortable	Warm	Hot	
•					January
•					February
•					March
	•				April
	•				May
	•				June
		•			July
	•				August
	•				September
	•				October
•					November
•					December

Add the checks

5	6	1		

Multiply by 8.33% for % of year

Heating	92
Comfortable	8

Energy-efficient glazings*

Bibek Bandyopadhyay
Director, Ministry of Non-conventional Energy Sources, Government of India

Windows are an essential architectural element. It establishes the link of the internal space of a building to the world outside and performs several functions, which include ventilation, controlling thermal insulation, and improving lighting within the building. However, at the same time, it provides the necessary barrier against the vagaries of nature. The successful design of a window depends on the optimization of these various factors including view and privacy. The magnitude of these functions vary in accordance with the climatic conditions. Also the type of building, its utilization pattern, the tradition, the social and cultural environment in which the building exists have profound influence on the above criteria. Windows, however, are poor thermal barriers. Conventional windows provide only a minimum level of thermal resistance. Heat gains and losses through windows have a large impact on the interior comfort level, energy use, and electric utility peak demand for the building. Windows are also sources of light. Therefore, design criteria for windows can provide distinct optical advantages and hence have implications on energy requirement of a building for lighting. The energy implications arising out of the above criteria are more important in the case of institutional and commercial buildings, which are often space-conditioned and functional mainly during daytime.

Glazings

A glazing is a transparent or translucent material used to cover the solar aperture. Conventional windows use clear glass as glazing. Essentially, glass sets the optical and thermal standards against which other glazing materials are judged. Glasses are partially transparent (typically 80% to 85%) throughout the near ultraviolet, visible and near infrared regions that compose the solar spectrum, but are practically opaque to longer-wave thermal radiation. Today, advanced glazings with innovative coatings and improved designs are available for energy conservation. These energy-efficient glazings minimize unwanted solar gains in summer and heat losses in winter while maximizing the amount of useful daylight in buildings. The performance characteristic parameters of a glazing from energy conservation point of view are defined in Box 1.

Box 1 Performance characteristic parameters of a glazing

Visible transmittance (T_v)	The fraction of visible light incident on the glazing that transmits to the interior
Shading coefficient (SC)	Describes the ability of a glazing to transmit solar heat gain relative to the solar heat gain of a 3 mm clear single glass pane
Coolness factor (K_e)	It is the ratio of the visible transmittance to the shading coefficient of a glazing, also known as glazing efficacy
Heat transfer coefficient (U)	Measure of thermal conductivity (W/m2K)

Multiple glazings

Windows may be constructed with multiple glazings (from two to three), depending on the climate type. These windows provide a heat barrier through insulating air cavity between glass layers and are known as IGUs (insulating glass units). The glasses are held apart by spacer bar and the air cavity contains a desiccant material.

Sun is the principal source of heat and light for a building. The average solar energy received outside the earth's atmosphere on a unit surface normal to sun's rays is 1367 W/m². This is known as solar constant. The solar energy when passes through the atmosphere gets diminished because of absorption and scattering. The amount of solar energy received per unit area on the surface of the earth is, therefore, of smaller magnitude. The energy of the sun is received as electromagnetic radiation with waves of varying lengths. The solar spectrum, which is received on the earth, extends from 290 nm to 3000 nm. It can broadly be divided in three wavelength regions (Table 1).

Solar control glazings

The primary objective of these glazings is to control the entry of solar heat and light into the buildings. These

---

*The views expressed in this article are those of the author in his individual capacity.

Table 1 Solar spectrum

Wavelength regions	Wavelength ranges	Share in total solar radiation
Ultraviolet radiation	290-400 nm	47%
Visible light VIBGYOR	400-770 nm	
Near infrared	770-3000 nm	53%

nm (nanometer) = 10^{-9} m

are essentially non-selective metallic films, which are deposited on glass or polymer substrates. Though these glazings are one of the earlier developments, but recently new material combinations and coating processes have contributed to the development of unique products. In these techniques, solar control is achieved either by absorptive process or by reflective process.

Heat absorbing glasses

The heat absorbing glass is basically an admixture of metallic oxides in its composition. Due to its wavelength filtering properties, which vary according to the quantity of the dopant and the thickness used, these glasses can substantially reduce the glare and excessive sunlight to enter into the building. The glasses are known as tinted glasses. The solar radiation absorbed, however, raises the temperature of the glass, which then radiates the thermal energy. This secondary emission contributes significantly to the heat gain of the interior of the building. This undesirable effect can be eliminated/reduced by adding an inner clear glass with the heat-absorbing glass. As the clear glass is opaque to long-wave radiation, it would prevent the heat radiation to flow to the interior of the building. Also the inner space between the two glasses would act as an insulating space offering resistance to heat conduction from the outside.

Heat-reflecting glasses

As the name suggests, these glasses are based on the reflective property of thin coatings, which are applied on the surface of the glass. The reflecting glasses have microscopically thin coating of pure metal like gold, silver or bronze between the two layers of glasses. The coating is applied on the cavity side of one of the double glazings of the window unit. Thus it provides low transmittance of solar energy with high insulating efficiency. Some glasses have a thin coating of metallic oxide projected on to the surface of the float glass as these leave the furnace. These glasses can considerably diminish the transfer of solar radiation (both heat and glare) in a hot climate.

Spectrally selective low-e glazings

Energy and light control characteristics are uniquely demonstrated by low-e (low emissivity) type of optical coatings. These optical coatings are normally applied to glass or plastic glazing materials. In order to understand the role of these optical coatings, we may consider two typical climate types—the cold and warm. In cold climate, an ideal window must be responsive to daylight and solar heat, so it might be transparent to the entire incident solar spectrum from 290 nm to 3000 nm. Also the window, in this climate, must offer maximum resistance to thermal radiation from inside the room. This would facilitate using all the solar energy available to heat the room and at the same time not allowing any heat to come out of the room. In warm climate, a window with glazing must prevent the infrared portion (770–3000 nm) of the solar spectrum to enter into the building. This is because any entry of heat to the building is undesirable but at the same time it must allow the visible light to come in so that daylight would be available through the window. These coatings essentially reduce the radiative heat transfer either from the outside of the building to the inside (in hot climate) or from the inside of the building to the outside (in cold climate) because of their low emissivity property. These glasses are comparatively new developments. However, they have already captured more than one-third market of glazing business in Japan, USA, and Europe. The performance characteristics of a few typical glazings mentioned above are given in Table 2.

Table 2 Performance characteristics of a few typical glazings

Type of glazing	Shading coefficient	Visible transmittance	Coolness factor	Heat transfer coefficient
Single glazing (clear glass)	1.00	0.91	0.91	6.46
Double glazing	0.82	0.78	0.95	3.31
Heat reflecting with plastic	0.56	0.74	1.32	2.50
Tinted glass (green)	0.72	0.75	1.04	6.45
Tinted glass (green) + clear glass	0.58	0.66	1.14	3.37
Low-e coating with glass	0.67	0.74	1.10	1.94
Low-e double glazing	0.85	0.78	0.92	2.27

Smart windows

In a varying climate, the requirement of heating and cooling of a building varies with seasons. Therefore, it would be desirable if the window glazings have a dynamic property rather than a static property as described above. Research is in progress in various parts of the world to devise such a window, which would automatically respond to the climate of the place. Such windows are known as 'smart windows'. Essentially these are chromogenic devices based on electrochromic, photochromic or thermochromic principles. Through thermochromic devices, changes in optical properties can be achieved when there is a change in temperature while such changes in photochromic devices are achieved with the amount of incident radiation. The optical properties of these devices vary from a high transmittance bleached state to a low transmittance coloured state depending on the input signal. The optical properties of the glazings based on electrochromic devices change when a current is applied. The electrochromic glazings hold the greatest potential for future commercial applications.

With use of suitable energy-efficient glazings, windows can be transformed from energy liability to energy assets. For a new construction, the use of such glazings would reduce the air-conditioning rating of the building and thereby related investment. This would not only offset the higher initial investment for the glazings being used but would also have a long-term consequence with respect to environmental benefits due to reduced energy consumption and use of lower amount of chlorofluorocarbons. The energy performance of a window with advanced glazings, however, will depend not only upon the size and type of the windows but also on specific weather parameters (such as solar radiation, temperature of the interior and exterior spaces, humidity, and windspeed) of the project site.

Appendix VI

Products and services offered by various companies

Engineering and construction

Janus Engineering (P) Ltd
K-55, Jangpura Extension
New Delhi – 110 014

Services
Lighting and electrical engineering

Confoss Constructions
E-108, Greater Kailash Enclave-I
New Delhi – 110 048

Services
Engineers, contractors, and builders

Equipment suppliers

Acoustic and thermal insulation

Bakelite Hylam Ltd
Regd off.: 7-2-1669, Sanat Nagar
PB No. 1908
Hyderabad – 500 018
Andhra Pradesh

Products
Decorative electrical switchboards, decorative laminates, paints, pre-laminated particle boards, thermal insulating material

Bharat Heat Insulation and Refractory Industries
Regd off.: 271, GIDC
Makarpura, Vadodara – 390 010
Gujarat
Tel. 574 3450/ 573 5816/ 578 8208

Products
Thermal and acoustic insulating material

Calcutta Polyurethanes Pvt. Ltd
Regd off.: 137
Biplabi Rash Behari Basu Road
III Floor (Canning Street)
Kolkata – 700 001, West Bengal

Products
Insulation material, polyurethane cast flooring

Caprihans India Ltd
'D' Block, Shiv Sagar Estate
Dr Annie Besant Road, Worli
Mumbai – 400 018, Maharashtra

Products
Decorative laminates, plasticized film, PVC products

Chauhan Insulations and Packaging
Regd off.: 50, Ponnappa Chetty Street
Chennai – 600 003
Tamil Nadu

Products
Thermal insulating material

India Gypsum Ltd
Regd off.: 815, Tolstoy House
8th Floor
15–17 Tolstoy Marg
New Delhi – 110 001

Products
Gypsum boards, partitions

India Rockwool Co. Ltd
Regd off.: 1, Ansari Road,
Darya Ganj
New Delhi – 110 002

Products
Thermal insulating material

Jolly Board Ltd
Regd off.: 501, Rewa Chambers
31, Marine Lines
Mumbai – 400 020
Maharashtra

Products
Acoustic materials, hardboards

Kitply Industry Ltd
Regd off.: Rungagora Road
Tinsukia – 786 125
Assam

Lloyd Insulations (India) Ltd
Punj House
M-13, Connaught Place
New Delhi – 110 001

Products
Chemicals for waterproofing, thermal insulation material

Air curtains

Air Devices Corporation
12/A, Bombay Shopping Centre
Race Course Road
Baroda – 390 005

Jet India
218, Allied Industrial Estate
Off. MMC Road Mahim (W)
Mumbai – 400 016

Air-flow meters

Air Devices Corporation
12/A, Bombay Shopping Centre
R C Dutta Road
Baroda – 390 005

J N Marshall Systems and Services
P.B. No. 1, Mumbai–Pune Road
Kasarwadi, Pune – 411 018

Toshniwal Sensors Pvt. Ltd
E/19-20,
Makhupura Industrial Area
Ajmer – 305 002

Techmark Engineers and Consultants
56 – D, DDA Flats,
Masjid Moth, Phase II,
New Delhi – 110048

Bricks and blocks

AEC Cements & Constructions Ltd
1 Floor, Electricity, Lal Darwaja
Ahmedabad – 380 001
Gujarat

Products
Pozzolana cement, flyash products

Asoka Brick Industries
Regd off.: 250, Poonamalee High Road,
Aminjikarai
Chennai – 600 029
Tamil Nadu

Products
Clay bricks

Bawa Enterprises
Ashoknagar
Mangalore – 575 006
Karnataka

Products
Bricks, burnt clay hollow blocks, ridges, tiles

Hindustan Tiles
3rd Street, Shukla Colony
Hinoo
Ranchi – 834 002
Bihar

Products
AC pipes, cement paints, cement waterproofing compound, concrete blocks, mosaic tiles

Hydratech Projects India Pvt. Ltd
8/3, Gandhi park,
Opposite Modi Hospital, Hauz Rani
New Delhi – 110 017

Products
Interlocking masonry blocks

Kalinga Concretes
Subodh KU Sahoo
Mansingh Patna
Cuttak
Orissa

Products
Fly ash bricks

Mechno Bricks Pvt. Ltd
307-308, Magnum House II,
B-3, Karampura Complex
New Delhi – 110 015

Products
Clay bricks and tiles

Nagarjuna Bricks
Ameena Complex,
Opposite Survey of India, Uppal
Hyderabad
Andhra Pradesh

Products
Hollow cement bricks and mosaic tiles

Nagson & Co.
4/1, Tumkur Road, Yeshwanthpur
Bangalore – 560 022
Karnataka

Products
Cement hollow blocks and bricks

Namha Fly-ash Brick Manufacturing & Construction Tech Pvt. Ltd
RS No. 44/1, Mattur post
Shimoga Taluk & Dist
Karnataka – 577 203

Products
Bricks and blocks

Orient Ceramics and Industries Ltd
Iris House, 16 Business Centre
Nangal Raya
New Delhi – 110 046

Products
Ceramic floor tiles in spectacular designs and shades

Rohini Flyash Brick Works
Plot No. 28, SIDCO Industrial Estate
Parvathipuram, Valadur
Cuddalore Dist
Chennai – 659 643
Tamil Nadu

Products
Flyash bricks and blocks

Sand Plast (India) Ltd
B-77, Raman Marg, Tilak Nagar,
Jaipur – 302 004
Rajasthan

Products
Bricks and tiles

Sitapur Plywood Manufacturers Ltd
Regd off.: Sitapur – 261 001
Uttar Pradesh

Products
Reconstituted wood products

Srinivasa Enterprises
Special Plot, Peenya I Stage, Near TVS Cross,
KSIDC Industrial Estate
Bangalore – 560 058

Products
Bricks and blocks

Suman Concrete Block Industries
No. 3/2/1, I Cross,
Behind Vijaynagar Telephone Exchange
MC Road, Vijaynagar
Bangalore – 560 079
Karnataka

Products
Cement bricks

Tara Nirman Kendra
Regd off.: B-32, Tara Crescent
Qutab Institutional Area
New Mehrauli Road
New Delhi – 110 016

Products
Bricks and blocks, ferrocement products, micro concrete roofing tiles

The Baliapatam Tile Works Ltd
PO Pappinisseri, Cannanore Dist
Pappiniserri - 670 561
Kerala

Products
Burnt clay products, processed rubber wood

Environment-friendly and energy-saving equipment

Auto Door Industries
C-37, Inderpuri
New Delhi – 110 012

Products
Aluminium doors, auto light, urinal flushers, gates-compound, rolling shutters, taps

Glass

Alankar Etched & Stained Glass Works
3/14, Mahalakshmi Mansion
I Main Road, Gandhi Nagar
Chennai – 600 020, Tamil Nadu

Products
Stained and Etched Glass

Continental Float Glass Ltd
III Floor, Pragati Kendra
Kapoorthala Commercial Complex
Aliganj, Lucknow – 226 020
Uttar Pradesh

Products
Float glass

Croda Polymers
1324, DB Gupta Road, Karol Bagh
New Delhi – 110 005

Products
Glass, Glass fibres, polymers

Fibro Style
1486, Wazir Nagar, Kotla Mubarakpur
(Near Hope Hall School)
Opp Defence Colony
New Delhi – 110 003

Products
FRP products, Wrinkle glass panels

FloatGlass India Ltd
104/108, Keshava, I Floor
Bandra–Kurla
Commercial Complex, Bandra (E)
Mumbai – 400 051

Products
Clear and tinted float glass

Gujarat Borosil Ltd
Khanna Construction House
44, Dr RG Thadani Marg
Worli, Mumbai – 400 018, Maharashtra

Products
Clear sheet glass

Gujarat Guardin Ltd
State Highway No. 13, Kondh Village
Bharuch Dist
Gujarat – 393 001

Products
Float glass

Gurind India Pvt. Ltd
60, Janpath
New Delhi – 110 001

Products
Glass products

Harrison Window
Onida Arcade
Opposite Mahaboob College
RP Road, Secundrabad – 500 003
Andhra Pradesh

Products
Glass

Mahaveer Mirror Industries
3, Devraj Mudali Street
Park Town
Chennai – 600 003, Tamil Nadu

Products
Heat reflective glass (exterior)

Marvel Glass Pvt. Ltd
'Utkarsh'
I Floor, Tithal road,
Valsad – 396 001, Gujarat

Products
Glass

Monsanto Enterprises Ltd
The Metropolitan Building
West Wing, 5th Floor
Bandra- Kurla Complex, Bandra (E)
Mumbai – 400 051

Products
Laminated architectural glass

Prime Time Industries
27, Shanti Industrial Estate
Ground Floor
Sarojini Naidu Road, Mulund (W)
Mumbai – 400 080

Products
Decorative and stained glass

The Indo-Asahi Glass Co. Ltd
3, Hungerford Street
Kolkata – 700 017, West Bengal

Products
Glass

Vijay Industrial Engineering Corporation
16/7/2B, Keyatalla Road, I Floor
Near Nazrulmanch
Southern Avenue
Kolkata – 700 029
West Bengal

Products
Metal products, stained glass, steel sections – light gauge

Vishwas Industries
663, Mallapa's New Market
Chickpet
Bangalore – 560 053
Karnataka

Products
FRP roofing sheets, glass

HVAC equipment suppliers

AC Humidification Engineers Pvt. Ltd
Regd office: A-33/29
Haneela Complex Guru Nanak Pura
Vikas Marg, Shakarpur
New Delhi – 110 092

Products
Air purifiers

Amtrex Appliances Ltd
Regd off.: "Rachna" PO Bombay Garage
Shahibaug Road
Ahmedabad – 380 004, Gujarat

Products
Air-conditioners

Arctic India Engineering Pvt. Ltd
20, Rajpur Road
Delhi – 110 054

Products
Air purifiers, Desiccant-based drying, energy recovery and cooling, indoor air environment control system.

Beacon Engineering (India) Ltd
Regd off.: 1206, Surya Kiran
19, KG Marg, New Delhi – 110 001

Products
Air curtains, elevators, insect killers

Blue Star Limited
Block 2-A, DLF Corporate Park
DLF Qutab Enclave, Phase –III
Gurgaon – 122 002

Products
Air-conditioning and refrigeration systems and products

Carrier Aircon Ltd
Kherki Daula Post, Nasingpur
Gurgaon – 122 001
Haryana

Products
Air conditioners and heaters

Caryaire Equipment's India Pvt. Ltd
A-10, Sector 59
Noida – 201 301

Products
Air distribution products for ceiling and wall application, double skin air treatment units using efficient fans, industrial/ commercial evaporative coolers, modular VAV systems from acutherm, USA, sound attenuators, centrifugal fans, inline fans, duct insulation material.

Fedders Lloyd Udyog Co. Ltd
Regd off.: Punj House
M-13-A Connaught Place
New Delhi – 110 001

Products
Air-conditioners

Godrej and Boyce Manufacturing Co. Ltd
Regd off.: Pirojshahnahar
Vikhroli
Mumbai – 400 079

Products
Air-conditioners, locks, panels and angles, partitions, doorknobs, doors, kitchen cabinets

Kirloskar Copeland Limited
1202/1, Ghole Road
Pune – 411 005

Products
Energy-efficient power Slash CR6 compressors for air-conditioning

LG Electronics India Pvt. Ltd
A-41
Mohan Co-operative Industrial Estate
Mathura Road, New Delhi

Products
Air conditioners

MEC India Ltd
Regd. Off. Kanchenjunga Building
18, Barakhamba Road
New Delhi – 100 001

Products
Air-conditioners

Suncontrol Airconditioning Co.
26/2
Shiv Krishna Daw Lane
Kolkata – 700 054

Products
Air-conditioners

Tecumseh Products India Ltd
Balanagar Township
Hyderabad – 500 037

Products
Energy-efficient compressor

Thermax Limited
Absorption Cooling Division
Chinchwad
Pune – 411 019

Videocon International Ltd
604, Bhikaji Cama Bhawan
Bhikaji Cama Place
New Delhi – 110 066

Products
Air-conditioners, water coolers

Voltas Limited
19, J N Heredia Marg
Ballard Estate
Mumbai – 400 038

Products
Voltas has a proven track record and expertise in the manufacture of room air-conditioners, industrial air-conditioning and refrigeration equipment

Infrared heating system

Litel Infrared System Pvt. Ltd
J-284, MIDC, Bhosari
Pune – 411 026

Lighting products

Ankur Lighting
2, Gagan Vihar
New Delhi – 110 051

Products
Lights and luminaires

Bajaj Electricals Limited
15/17, Sant Savta Marg
Reay Road
Mumbai – 400 010

Crompton Greaves Limited
(Lighting Division)
Dr E Morses Road
Worli
Mumbai – 400 018

Decon Lighting Pvt. Ltd
5, Lok Nayak Bhawan, Khan Market
New Delhi – 110 003

Products
Luminaires and accessories

Dyna Lamp & Glass Works Ltd
35, Nungambakkam High Road
Pottipati Plaza, III Floor
Chennai – 600 034, Tamil Nadu

Products
Lamp

ECE Industries Ltd
ECE house
28A, Kasturba Gandhi Marg
New Delhi – 110 001

Products
Electric meters, elevators, lights, power and distribution transformers, resin cast dry type distribution transformers, switchgear

Elemech Group
E-2-3-4, Manish Nagar
Jay Prakash Road, Andheri (W)
Mumbai – 400 053

Products
Electrical installations

GE Lighting
Maker Chambers III, 1st Floor
Jamanalal Bajaj Road
New Delhi – 110 015

Indo Asian Fusegear Ltd
Lighting Div.
A-39, Hoisery Complex
Phase-II, Noida
Distt. G.B. Nagar
Uttar Pradesh – 201 305
India

Products
CFL lamps, luminaires, and controls gears.

Legrand Luminaries Pvt. Ltd
C-122, Naraina Vihar
New Delhi – 110 028

Products
Chokes, luminaires

Osram India Pvt. Ltd
1/95, Market Road,
Bhai Veer Singh Marg
New Delhi – 110 001

Products
Lamps and chokes

Philips India Ltd
P-65, Taratolla Road
Kolkata – 700 088

Products
CFL lamps, lumanaires and control gears

Surya Roshni Ltd
Padma Tower – I
Rajendra Place
New Delhi – 110 008

Products
Luminaires, steel products

Sylvania and Laxman Ltd
68/1-3, Najafgarh Road
PO Box 6224
New Delhi – 110 015

Products
Electric fittings and accessories, lights/luminaires

Wipro Lighting
Tulsi Chambers, Opposite St. Francis School,
Jalna Road, Aurangabad – 461 001
Maharashtra

Products
Electric accessories, luminaires

Lux meter (for illuminance levels)

Conin Prakriti Instrumentation
16, Rajendranagar Industrial Estate
Post Mohan Nagar
Ghaziabad 201 007, Uttar Pradesh

Research Instrumentation
A-10
Naraina Industrial Area, Phase – I
New Delhi – 110 028

Occupancy sensors / controls / building automation

Building automations and control systems

Johnson Controls (India) Pvt. Ltd
B-37, RV House
Veera Desai Road
Off Link Road, Andheri (W)
Mumbai – 400 053

Products
Building control systems

Philips India Ltd
M 38-1 Middle Circle
Connaught Place
New Delhi – 110 001

Pyrotech Marketing and Projects Pvt. Ltd
412, Shakuntala Apartments
59, Nehru Place
New Delhi – 110 019

Shivalika Proenergetics Ltd
A-116, Madhuvan, Vikas Marg
New Delhi – 110092

Tata Honeywell Ltd
55-A/8 & 9, Hadapsar Ind. Estate
Pune – 411 013

Products
Heating ventilation and air conditioning control system, lighting control system, elevator control system, sanitary and plumbing control system, electrical monitoring and control system.

Tata Honeywell Ltd
917, International trade Tower
Nehru Place, New Delhi – 110 019

Relative humidity meters

Aimil Ltd
Naimex House, A-8
Mohan Co-operative Industrial Estate
Mathura Road
New Delhi – 110 044

Digital promoters (India) Pvt. Ltd
505, Vishal Bhawan
95, Nehru Place
New Delhi – 110 019

Instrument Research Associates
PB No. 2304, 228
Magadi road
Bangalore

OPTEL
104, Brahmanwadi
K A Subramanian Road
Matunga
Mumbai – 400 019

Recuperator for waste heat recovery

Encon Thermal Engineers (P) Ltd
308, Race Cource towers
Race Cource Circle
Baroda – 390 015

Soft starters with energy saving feature

Allen-Bradley India Ltd
C-11, Industrial Area
Site - 4 Sahibabad
Ghaziabad – 201 010

Bharat Bijlee Ltd
Industrial Electronics Division
501-502, Swastik Chambers
Chembur, Mumbai – 400 023

Crompton Greaves Ltd
Industrial Electronic Division
71/72, MIDC industrial Area
Satpur, Nashik – 422 007

Jeltron Systems (India) Pvt. Ltd
Bag No. 49, 6-3-99/2
Vaman Nayak Lane, Umanagar Colony
Begumpet, Hyderabad – 500 018

Siemens Limited (MDA Dept.)
Electric Mansion
1986, Appasaheb Marathe Marg
PB No. 1911, Prabhadevi
Mumbai – 400 025

Solar photovoltaic and solar water heating systems

Aditi Impex Ltd
G-13 Shreeji Palace
PO Navjivan, Ahmedabad – 380 014

Products
Importers of PV modules, solar water pumps, trackers, inverters and products pertaining to solar energy, PV module mounting systems.

Aditya Online
No. 441, 8th Cross
Mahalakshmi Layout
Bangalore, India – 560 086

Products
Solar garden lights, solar electric power systems, solar outdoor lighting systems, solar water pumping systems, Solar Lanterns, Solar Lanterns with AM/FM radio, Solar Caps, Solar Fan.

AJ Electronics
Prerana, 21 Amar Society
44/2, Erandwane, Pune – 411 004

Amar Urja Ltd
12-Shivpuri, Bulandshahar,
India – 203 001

Products
Photovoltaic systems, solar outdoor lighting systems, solar electric power systems, solar garden lights, DC to AC power inverters, solar roofing systems.

Ammini Energy Systems Pvt. Ltd
Industrial Estate Pappanamcode,
Thiruvananthapuram, India – 695 019

Product
Compact fluorescent lighting fixtures and ballasts, photovoltaic systems, solar outdoor lighting systems, solar garden lights, remote home power systems, solar lantern, solar home-lighting systems

Ankur Scientific Energy Technologies (P) Ltd
'Ankur', Near Old Sama Jakata Naka
Vadodara - 390 088

Product
Solar water heating system

Auroville Energy Products
Auroville – 605 101
India

Products
Solar electric power systems, photovoltaic systems, photovoltaic modules, wind energy systems (small), wind turbines (small), renewable energy system batteries.

Bharat Heavy Electricals Limited (BHEL)
Siri Fort Road
New Delhi, India

Products
PV modules, solar water heating systems, wind energy systems (small), solar outdoor lighting systems.

Britto Energy Engineers
34, Daattani Trade Centre
Chandawarkar Road
Opposite Railway Station
Borivili (West)
Mumbai – 400 092

Product
Solar water heating system

CASE Pvt. Ltd
458, Udyog Vihar, Phase V
Gurgaon – 122 016

Products
solar water heaters, solar water pumping systems, solar cooking products.

Centre for Energy Initiatives
No. 6, 7 and 8, IVth 'N'
Block, Rajajinagar Entrance
Dr Rajkumar Road
Bangalore – 560 079

Products
Solar water heating systems, energy efficient appliances, biomass energy systems, remote home power systems, solar electric power systems, hydro-electric turbines (small).

Deep Engineers
C-4/132, Secor-6, Rohini
Delhi – 110 085

Products
Solar photovoltaic system

EcoSolar Systems (India) Ltd
17/A/2 Pune–Sinhgad Road
Pune – 411 030

Products
Photovoltaic cells, photovoltaic modules, battery chargers, hydro energy systems (small <50 kW), solar air heating systems, solar electric power systems, solar water heating systems, solar PV lanterns and other solar PV and thermal applications.

EcoSolar Systems (India) Ltd
117/A/2 Pune–Sinhgad Road,
Pune – 411030, India

Products
Photovoltaic cells, photovoltaic modules, battery chargers, hydro energy systems (small <50 kW), solar air heating systems, solar electric power systems, solar water heating systems, solar PV lanterns and other solar PV and thermal applications

Emvee Solar Systems
No 253, 5th Main Road, 10th Cross
Vasanthappa Block
Bangalore – 560 032

Product
Solar water heating system

Ensemble Systems
6/6, Sinchan Nager
Pune – 411 020, India

Products
solar water heaters, custom solar energy systems.

ERENA Technologies (India)
No. 52, Dodanna Industrial Estate
Near Peenya 2nd stage
Bangalore – 560 091, India

Products
Solar air heating systems, solar cooking systems, solar water heating systems, solar tracking systems, hydro energy system components (large), biomass energy systems

Enertech Marketing Services
417, 41st Cross
Jayanagar 5th Block
Bangalore, India – 560 041

Products
Photovoltaic modules, solar electric power systems, solar outdoor lighting systems, solar water heating systems.

Flexitron
114, Kathalipalya
6th Floor, Vi Block, Koramangala
Bangalore – 560 034

Product
Solar photovoltaic system

Gitanjali Solar Enterprises
P/14, Kasba Industrial Estate
Phase I, Em Bye Pass
Kolkata – 700 054

Product
Solar Photovoltaic system

Hitech Engg. Works
4-5/B, Ancillary Indl. Area
P O Hatia
Ranchi – 834 003

Product
Solar photovoltaic system

Hot Hold Systems Pvt. LTD
474 1st floor 1st main 3rd Block
Basaveshwar Nagar, Bangalore,
India – 560 079

Product
Solar water heaters, solar cooking products

Ibex Electronics
175/54, Tathna Vilasa Road
Bangalore – 560 004

Products
Solar photovoltaic system

Jain Irrigation Systems India Ltd
Jain Plastic Park
PO 72, Jalgaon – 425 001

Products
'Jain Sun Watt' Solar Water Heating System; PVC/PC Sheets for roofing, domes, interiors, furniture's, false ceilings, partitions etc.; plastic foot valve used for lifting water from open-wells

Jash Projects Pvt. Ltd
30/384 Telang Road
Mumbai – 400 019

Products
Solar water heaters, solar pool heating systems, waste treatment systems, energy efficient lighting, photovoltaic systems (PV systems, solar electric systems)

Kaynes Energy Systems
23-25
Belagola Food Industrial Estate Metagalli
Mysore, Karnataka – 517 016

Products
Photovoltaic modules, photovoltaic systems, solar outdoor lighting systems, solar lanterns, solar domestic lighting systems, solar water heating systems, solar garden lights

Kaushal Solar Equipment (P) Ltd
767/7, Neel-Kusum
Deccan Gymkana, Pune
Mississippi India – 411 004

Products
Photovoltaic systems, solar electric power systems, solar garden lights, solar pool heating systems, solar water heating systems, solar water pumping systems, solar lanterns, solar garden lights, solar PV mini-kits wind-PV hybrid systems

Kaushal Solar Equipment (P) Ltd
767/7, Neel-Kusum
Deccan Gymkana, Pune
Mississippi India – 411 004

Products
Photovoltaic systems, solar electric power systems, solar garden lights, solar pool heating systems, solar water heating systems, solar water pumping systems, solar lanterns, solar garden lights, solar PV mini-kits, wind-PV hybrid systems.

Kotak Urja Private Limited
378, 10th Cross, IV - phase
Peenya Industrial Estate
Bangalore India – 560 058

Products
Solar water heating systems, solar air heating systems, solar outdoor lighting systems, solar pool heating systems, solar garden lights, solar crop dryers, solar safe drinking water systems, solar home-lighting systems

Machincraft (Pune) Pvt. Ltd
15/4A, Vasudeo Estate
Opposite Shankar Maharaj Temple
Pune Satar Road, Pune – 411 043

Product
Solar Photovoltaic system

Navkar Energy Systems Pvt. Ltd
NR: Guptanagar Bus Stand
Guptanagar, Vasna
Ahmedabad, India – 380 007

Products
Solar water heating systems, solar pool heating system components, tankless water heating systems, solar air heating systems, solar water heating components, solar water heating systems

Neptro Renewable Energy (India) Ltd
3rd Floor, Gupta Towers
50/1, Residency Road, 1st Cross
Bangalore – 560 025

Product
Solar water heating system

Novel Energy (P) Ltd
550, Mandakini Enclave, Kalkaji
New Delhi – 110 019

Product
Solar water heating system

Planters Energy Network
171/2, MK University Road
Rajambadi, Madurai – 625 021

Product
Solar water heating system

Prabhu Energy Systems (P) Ltd
Product types: solar water heaters
Address: Hameed Complex Alake
Mangalore, India – 575 003

Prabhu Energy Systems
2nd Floor, Hammed Complex
Kuloor Ferry Road
Mangalore – 575 003

Product
Solar water heating system

Premier Solar Systems (P) Ltd
41 & 42, Sri Venkateswara
Co-operative Indl. Estate, Balanagar
Hyderabad – 500 037

Product
Solar photovoltaic system

Professional Lighting Pvt. Ltd
25, Singh Ind Estate No 3, Ram Mandir Rd, Goregaon (W), Mumbai
Mumbai – 400 104

Products
Back-up power systems, DC lighting, portable power systems, solar outdoor lighting systems, UPS, solar electric power systems, emergency photoluminescent wayfinding systems.

Punjab Power Packs Ltd
B-98, Phase VIII
SAS Nagar (Mohali)
Chandigarh – 160 059

Product
Solar photovoltaic system

Rajasthan Electronics & Instruments Limited
2, Kanakpura Industrial Area
Sirsi Road, Jaipur – 302 012

Products
Photovoltaic modules, photovoltaic systems, solar water pumping systems, solar electric power systems, cathodic protection systems, solar lantern

Reliant Solar Systems Pvt. Ltd
54, Vinobha Puri (Basement)
Lajpat Nagar
New Delhi – 110 024

Product
Solar water heating systems, solar cooking systems, solar pool heating systems, tankless water heating systems, solar air heating systems, solar garden lights, solar lanterns, photo voltaic panels

Renewable Energy Systems Pvt. Ltd
D-52, Phase V
Industrial Development Area
Jeedimetla
Hyderabad – 500 855

Product
Solar photovoltaic system

Ritika Systems Pvt. Ltd
B-279, Okhla Industrial Area
Phase I
New Delhi – 110 020

Products
Solar photovoltaic system

Savemax Solar Systems Pvt. Ltd
Jayaprabha, Jadhavnagar
Vadgaon BK, Pune – 411 041

Product
Solar flat plate collectors, solar water heating systems, solar cookers, solar lanterns, solar homelighting systems

Shiv Shakti Electronics Pvt. Ltd
F-274, Flatted Factories Complex
Okhla Industrial Area
New Delhi – 110 020

Product
Solar photovoltaic system

Simmark
Survey No. 2/1, Ghorpuri
Pune – 411 001

Products
Air-heating system components, renewable energy system batteries, solar air heating systems, solar pool heating systems, solar water heating systems.

SD Solar Systems India Pvt. Ltd
11, Shahajanabad
Bhopal – 462 001

Product
Solar water heating system

Soladur Energy Systems Pvt. Ltd
6-3-354/15, Hindi Nagar
Banjara Hills
Hyderabad – 500 034

Product
Solar water heating system

Solartech
SU 1557, Sector 7-7
Chandigarh – 160 019

Product
Solar water heating system

Solarshopee
Raghubar Niwas
Gurudwara Colony
Lohegaon, Pune, India – 411 047

Products
Solar air heating systems, solar cooking systems, solar electric power systems, solar outdoor lighting systems, solar pool heating systems, solar water heating systems

Solker Industries Limited
37, Jayalakshmipuram
I st. Nungambakkam High Road
Chennai – 600 034

Product
Solar Photovoltaic system

Solchrome Systems India Limited
61, Sector-5, Parwanoo
India – 173 220

Products
Solar water heating systems, solar water heating components.

Sudharsan Industries
5, Jijamata Colony
Aurangabad – 431 001

Product
Solar water heating system

Sun Power
Suit 2E, 52B, Shakespeare Sarani
Kolkata – 700 017

Product
Solar Photovoltaic system

Sunlit Enterprises
Visva-Sudha
9, Shamsundar Hsg Soc
Near Mhatre Bridge
Pune – 411 030

Product
Solar water heating system

Suryodaya HI-Tech Engineering Pvt. Ltd
1-10-68/4, Chikoti Gardens
Begumpet
Secundrabad – 500 016

Product
Solar water heating system

Su Solartech Systems (P) Ltd
SCO 184, Opposite Sports Complex
Sector 7-C, Chandigarh – 160 019

Sunbeam Solar Thermal Co.
1114/11 University road, Pune
India – 411 016

Product
Solar water heater, solar cooker

Sunline Solar Systems Pvt. Ltd
1836, Sadashiv Peth
Deshmukh Wadi, Pune – 411030

Products
Solar water heating systems, solar pool heating systems.

Suntime Energy Ltd
E-3 Lajpat Nagar
New Delhi, India

Products
Solar water heating systems, solar electric power systems, photovoltaic modules, solar water heating components, solar cooking system

Surya Shakti
739, Industrial Area Phase-II
Chandigarh, India – 160 019

Products
Solar water heating systems, solar air heating systems, solar cooking systems, solar pool heating system components.

Suryodaya Hi-Tech Engineering (P) Ltd
1-10-68/4, Chikoti Gardens
Begumpet
India – 500 016

Products
Solar water heating systems, solar water heating components, solar flat plate collectors, solar parabolic trough collectors, solar selective coated fins, solar electric power systems, solar air heating system components, solar garden lights, solar water heating systems, wind turbines (small)

Tata BP Solar India Ltd
Plot No. 78, Electronic City
Hosur Road
Bangalore – 561 229

Products
Solar PV and solar thermal systems

The Energy Pool
34, Friends Colony, Chamba Ghat
Solan – 173 213

Products
Solar photovoltaic system

Thermosol Equipment
55-G, (NP_, SIDCO Industrial Estate
Ambattir, Chennai – 600 098

Product
Solar water heating system

Udhaya Semiconductor (P) Ltd
1/482, Avanashi road
Neelambur
Coimbatore – 641 014

Product
Solar Photovoltaic system

UP Hill Electronics Corporation Ltd
A-1/26, Vishwas Khand II
Gomati Nagar
Lucknow – 226 010

Product
Solar Photovoltaic system

Webel-SL Energy Systems Ltd
Plot No N1, Block -GP, Sector V
Salt Lake Electronics Complex
Kolkata – 700 091

Product
Solar photovoltaic system

Sun control films

Garware Plastics and Polyster Ltd
Western Express Highway
Vile Parle
Mumbai – 400 057

Temperature monitors / controllers / sensors

Aimil Ltd
Naimex House, A-8
Mohan Co-operative Industrial Estate
Mathura Road, New Delhi – 110 044

Digital Promoters (India) Pvt. Ltd
505, Vishal Bhawan
95, Nehru Place, New Delhi – 110019

Instronix India Pvt. Ltd
3E / 12, Jhandewalan Extension
New Delhi – 110 055

Masibus Process Instruments (P) Ltd
B/30, GIDC Electronic Zone
Gandhinagar – 382 044

Toshniwal Sensors Pvt. Ltd
E/19-20, Makhupura industrial Area
Ajmer – 305 002

Time / control switches (timers)

Escol Electromech Pvt. Ltd
Kripalani Estate
Saki Vihar Road
Mumbai – 400 072

Indo Asian Marketing Ltd
A-59, Okhla Industrial Area
Phase II
New Delhi – 110 020

Jayshree Enterprises
(Marketing Division)
101, Prabodhan Apartment
64/9, Erandwane
Pune – 411 004

Larsen & Toubro Limited
PO Box No. 8901, Powai
Mumbai – 400 072

MDS Switchgear Limited
314-317, Shah & Nahar Road
Off. Dr E Moses Road
Worli
Mumbai – 400 018

Products
Time switches

Variable speed drives

Allen Bradley India limited
A-5, Second Floor, Kailash Colony
New Delhi – 110048

Asean Brown Boveri limited
Guru nanak Foundation building
15–16, Qutub Institutional Area
New Delhi – 110 067

Crompton Greaves Limited
Industrial Electronic Division
71/72, MIDC industrial Area
Satpur, Nashik – 422 007

L&T Static Control
(Electronic Controls Division)
Larsen & Toubro Ltd
PO Box No. 8901, Powai
Mumbai – 400 072

Jeltron Systems (India) Pvt. Ltd
Bag No. 49, 6-3-99/2,
Vaman Nayak Lane, Umanagar Colony
Begumpet, Hyderabad – 500 018

Kirloskar Electric Company Ltd
Unit IV, Belawadi Industrial Area
Mysore – 570 005

Siemens Limited (MDA Dept.)
Electric Mansion
1986, Appasaheb Marathe Marg
PB No. 1911, Prabhadevi
Mumbai – 400 025

The General Electric Company of India Ltd
Electronics Unit, DA 21
Salt Lake City
Kolkata – 700 064

Index

A

absorption chillers, RETREAT, Gurgaon **115**
absorption technology, TCI Ltd, Gurgaon **72**
active solar interventions, application of **21**
Ahmedabad
    Mahendra Patel **161**
    Sangath **151**
    TRC **155**
AIIS (*see* American Institute of Indian Studies)
air
    cavities within walls **8**
    distribution, TCI Ltd, Gurgaon **73**
    panels for space heating, HPSCB, Shimla **26**
air-conditioning
    monitoring system for TCI Ltd, Gurgaon **74**
    property at Civil Lines, Delhi **83**
air-handling units, TCI Ltd, Gurgaon **73**
air-heating system, roof-based **12**
Airport and staff housing colony, Kargil **45**
airport staff housing, Kargil **47**
ambient temperature, control of, IREP TC, Delhi **88**
American Institute of Indian Studies (AIIS), Gurgaon **138**
architectural design, Vikas, Auroville **206**
architectural integration, effect of, Sri Aurobindo Ashram, New Delhi **95**
architectural interventions **1**
artificial illumination
    HP State Co-operative Bank, Shimla **28**
    TCI Ltd, Gurgaon **73**
    SEC, Gurgaon **130**
Auroville
    Kindergarten School **214**
    La Cuisine Solar **210**
    Vikas Apartments **206**
    Visitors' Centre **218**

B

Bangalore
    Mary Mathew **173**
    TERI office building-cum-guest house **177**
Baptist Church, Chandigarh **124**
Bhopal, WALMI **119**
Bidani House, Faridabad **66**
bioclimatic approach to design, IREP TC, Delhi **86**
bioclimatic architecture, example of, WALMI, Bhopal **119**
building
    and system design, TERI, Bangalore **180**
    automation system, Mahendra Patel, Ahmedabad **164**
    behaviour, scientific observation of, TRC, Ahmedabad **159**
    design, Nisha, Goa **185**
    envelope **4**
    fabric, SOS TCV, Dehradun **78**
    materials, energy contents of **6**
    orientation **4**
    plans, SOS TCV, Dehradun **77**
    systems, CMC House, Mumbai **224**
    technologies, Vikas, Auroville **207**

C

ceiling lights, control of, TCI Ltd, Gurgaon **73**
central evaporative cooling, Neelam and Ashok Saigal, Gurgaon **104**
Chandigarh
    Baptist Church **124**
    PEDA office complex **62**
climate in Shimla **21**
climate-responsive building form, PEDA, Chandigarh **63**
climate-responsive design, Bidani house, Faridabad **66**
climate-responsive device, property at Civil Lines, Delhi **82**
climate, Chandigarh **62**
climatic response, Bangalore **173**
CMC (*see* Computer Maintenance Corporation House, Mumbai)
Computer Maintenance Corporation (CMC) House, Mumbai **223**
concept of
    MLA Hostel, Shimla **34**
    SOS TCV, Dehradun **76**
construction techniques **5**
    Kindergarten School, Auroville **215**
    Nisha, Goa **187**
    Visitors' Centre, Auroville **220**
conventional evaporative cooler, property at Civil Lines, Delhi **83**
courtyard planning, AIIS, Gurgaon **139**

D

daylight
    distribution, Himurja, Shimla **22, 23**
    integration, WBREDA, Kolkata **192**
daylighting
    and cross-ventilation, Baptist Church, Chandigarh **125**
    and heating, Himurja, Shimla **22**
    and ventilation, Visitors' Centre, Auroville **219**
    design, TERI, Bangalore **179**
    effect of **17**
    HP SCB, Shimla **28**
    Mary Mathew, Bangalore **175**
    SEC, Gurgaon **130**
DC HCC (*see* Degree College and Hill Council Complex, Leh)
DC HCC, the site for **41**
Degree College and Hill Council Complex (DC HCC), Leh **41**
Dehradun, SOS TCV **76**
Delhi, residence of Sudha and Atam Kumar **97**
design
    concepts **1**
    details, CMC House, Mumbai **224**
    elements **2**
    features, AIIS, Gurgaon **142**
design features
    Airport and staff housing colony, Kargil **46, 48**
    Baptist Church, Chandigarh **127**
    Bidani house, Faridabad **68**
    CMC House, Mumbai **227**
    DCHCC, Leh **44**
    Dilwara Bagh, Gurgaon **109**
    Himurja, Shimla **24**
    HP SCB, Shimla **28**
    IIHMR, Jaipur **149**
    IREP TC, Delhi **90**
    Kindergarten School, Auroville **217**
    LEDeG Trainees' Hostel, Leh **54**
    Madhu and Anirudh, Panchkula **61**
    Mahendra Patel, Ahmedabad **164**
    Mary Mathew, Bangalore **176**
    MLA Hostel, Shimla **37**
    Neelam and Ashok Saigal, Gurgaon **105**
    Nisha, Goa **188**
    NMC CHS, Gurgaon **136**
    PEDA, Chandigarh **65**
    property at Civil Lines, Delhi **84**
    RETREAT, Gurgaon **117**
    Sangath, Ahmedabad **154**
    SEC, Gurgaon **132**
    Silent Valley, Kalasa **205**
    SOS TCV, Dehradun **78**
    SPH, Jodhpur **169**
    Sri Aurobindo Ashram, New Delhi **96**
    Sudha and Atam Kumar, Delhi **100**
    Tabo Gompa, Spiti **56**
    TCI Ltd, Gurgaon **75**
    TERI, Bangalore **181**
    TRC, Ahmedabad **159**
    Vikas, Auroville **208**
    Visitors' Centre, Auroville **222**
    WBPCB, Kolkata **200**
    WBREDA, Kolkata **194**
    WALMI, Bhopal **123**
design optimization, WBPCB, Kolkata **197**

design response
  PEDA, Chandigarh 62
  TERI, Bangalore 178
design strategies
  WBREDA, Kolkata 190
  TCI Ltd, Gurgaon 69
design
  SPH, Jodhpur 166
  Sangath, Ahmedabad 152
Dilwara Bagh (*see* Dilwara Bagh, Country House for Reena and Ravi Nath, Gurgaon)
Dilwara Bagh, Country House for Reena and Ravi Nath, Gurgaon 106
direct heat gain 10, 11
dust and cobweb control, TRC, Ahmedabad 157

E

earth
  air tunnels 15
  contact, AIIS, Gurgaon 139
  for cooling, exploitation of, IREP TC, Delhi 88
efficient utilization of energy, RETREAT, Gurgaon 111
energy, efficient utilization of, RETREAT, Gurgaon 111
embodied energy, Mary Mathew, Bangalore 174
energy conservation
  CMC House, Mumbai 226
  IIHMR, Jaipur 149
energy consumption, HP SCB, Shimla 25
energy efficiency in architecture 1
energy-efficient design
  Mahendra Patel, Ahmedabad 161
  NMC CHS, Gurgaon 135
energy-efficient glazing system, write-up on 9
energy-efficient lighting
  design, WBREDA, Kolkata 192
  systems 1
  Sudha and Atam Kumar, Delhi 100
  RETREAT, Gurgaon 115
  WBPCB, Kolkata 199
energy gains, TCI Ltd, Gurgaon 74
energy-saving features
  MLA Hostel, Shimla 34
  TCI Ltd, Gurgaon 70
  WBPCB, Kolkata 196
energy saving strategies, HP SCB, Shimla 25
evaporating cooling 16
evaporative cooler, property at Civil Lines, Delhi 83

F

Faridabad, Bidani House 66
fenestration 4

fenestration and shading 8
  AIIS, Gurgaon 140
fenestration design 8
  WBPCB, Kolkata 197
fenestration pattern, IIHMR, Jaipur 148
fountain court, TCI Ltd, Gurgaon 71

G

garbage disposal, NMC CHS, Gurgaon 136
glazing systems 9
  WBREDA, Kolkata 193
Goa, Nisha's Play School 185
guest house building, SEC, Gurgaon 131
Gurgaon
  AIIS 138
  Dilwara Bagh 106
  NMC CHS 134
  residence for Neelam and Ashok Saigal 102
  RETREAT 111
  SEC 128
  TCI Ltd 69

H

heat
  collecting wall, HPSCB, Shimla 26
  condition, IREP TC, Delhi 87
  gain, distribution of, Himurja, Shimla 22
  transfer, TCI Ltd, Gurgaon 71
Hill Council Complex, site for 43
Himachal Pradesh State Co-operative Bank (HPSCB), Shimla 25
Himurja office building, Shimla 21
hot water system, Sri Aurobindo Ashram, New Delhi 93
HP SCB (*see* Himachal Pradesh State Co-operative Bank, Shimla)

I

IIHMR (*see* Indian Institute of Health Management Research, Jaipur)
illumination
  monitoring system for, TCI Ltd, Gurgaon 74
  source of, TCI Ltd, Gurgaon 73
Indian Institute of Health Management Research (IIHMR), Jaipur 145
indirect/diffused light, Sangath, Ahmedabad 153
insulation
  property at Civil Lines, Delhi 81
  and shading, Dilwara Bagh, Gurgaon 108
  and window design, Himurja, Shimla 23
  of the roof, IREP TC, Delhi 88
  CMC House, Mumbai 226

Mahendra Patel, Ahmedabad 162
Neelam and Ashok Saigal, Gurgaon 104
SEC, Gurgaon 130
Sudha and Atam Kumar, Delhi 98
WBREDA, Kolkata 192
Integrated Rural Energy Programme Training Centre (IREP TC), Delhi 86
internal heat build-up, DCHCC, Leh 44
IREP TC (*see* Integrated Rural Energy Programme Training Centre, Delhi)

J

Jaipur, IIHMR 145
Jodhpur, SPH 166

K

Kalasa, Silent Valley 201
Kargil, Airport and staff housing colony 45
Kindergarten School, Auroville 214
Kolkata
  West Bengal Pollution Control Board 195
  West Bengal Renewable Energy Development Agency 189

L

La Cuisine Solar, Auroville 210
landscaping 2
  WBREDA, Kolkata 191
LEDeG Trainees' hostel, Leh 49
Leh
  DC HCC 41
  LEDeG Trainees' hostel 49
lighting system, integration of daylighting with, CMC House, Mumbai 225
louvres 9

M

Mary Mathew, residence of, Bangalore 173
materials of construction, Tabo Gompa, Spiti 56
methods of construction, Tabo Gompa, Spiti 56
microclimate, control of, IREP TC, Delhi 88
MLA hostel, Shimla 33
Mumbai, Computer Maintenance Corporation House 223

N

Nainital, residence of Mohini Mullick 30
National Media Centre Co-operative Housing Scheme, Gurgaon (NMC CHS) 134

New Delhi, Sri Aurobindo Ashram 91
Nisha's play school, Goa 185
NMC CHS (*see* National Media Centre Co-operative Housing Scheme, Gurgaon)

O

outdoor spaces, SOS TCV, Dehradun 77
overhangs 9

P

Panchkula, residence for Madhu and Anirudh 59
passive architectural interventions, Neelam and Ashok Saigal, Gurgaon 102
passive architectural principles
    adoption of, Dilwara Bagh, Gurgaon 106
    IIHMR, Jaipur 145
passive architectural techniques 2
    Sudha and Atam Kumar, Delhi 97
passive condition techniques, IREP TC, Delhi 86
passive cooling system, TRC, Ahmedabad 156
passive cooling techniques 13
passive designing for load reduction, RETREAT, Gurgaon 112
passive downdraught cooling 16
passive heating techniques 10
passive solar concepts
    IREP TC, Delhi 87
    La Cuisine Solaire, Auroville 210
passive solar features
    DCHCC, Leh 42, 43
    Kindergarten School, Auroville 214
    Madhu and Anirudh, Panchkula 60
    Property at Civil Lines, Delhi 80
    SEC, Gurgaon 128, 131
    SEC, Gurgaon 130
    SPH, Jodhpur 167
    Tabo Gompa, Spiti 55
passive solar interventions, application of 21
passive solar techniques 2
    AIIS, Gurgaon 139
    SEC, Gurgaon 128
    WALMI, Bhopal 119
passive ventilation techniques, TERI, Bangalore 178
PEDA office complex, Chandigarh 62
performance
    of Sangath, Ahmedabad 153
    of SPH, Jodhpur 168
    survey, TCI Ltd, Gurgaon 74
    property at Civil Lines, Delhi 83
photovoltaic-gasifier hybrid power plant, RETREAT, Gurgaon 113

project details
    AIIS, Gurgaon 142
    Baptist Church, Chandigarh 127
    Bidani house, Faridabad 68
    CMC House, Mumbai 227
    DCHCC, Leh 42, 44
    Dilwara Bagh, Gurgaon 109
    Himurja, Shimla 24
    HP SCB, Shimla 28
    IIHMR, Jaipur 149
    IREP TC, Delhi 90
    Kindergarten School, Auroville 217
    La Cuisine Solaire, Auroville 212
    LEDeG Trainees' hostel, Leh 53
    Madhu and Anirudh, Panchkula 61
    Mahendra Patel, Ahmedabad 164
    Mary Mathew, Bangalore 176
    MLA Hostel, Shimla 37
    Neelam and Ashok Saigal, Gurgaon 105
    Nisha, Goa 188
    NMC CHS, Gurgaon 136
    PEDA, Chandigarh 65
    property at Civil Lines, Delhi 84
    RETREAT, Gurgaon 116
    Sangath, Ahmedabad 154
    SEC, Gurgaon 132
    Silent Valley, Kalasa 205
    SOS TCV, Dehradun 78
    SPH, Jodhpur 168
    Sri Aurobindo Ashram, New Delhi 95
    Sudha and Atam Kumar, Delhi 100
    Tabo Gompa, Spiti 56
    TCI Ltd, Gurgaon 75
    TERI, Bangalore 181
    TRC, Ahmedabad 159
    Vikas, Auroville 208
    Visitors' Centre, Auroville 222
    WBPCB, Kolkata 200
    WBREDA, Kolkata 194
    WALMI, Bhopal 123
property at civil lines, Delhi 80

R

rainwater harvesting
    NMC CHS, Gurgaon 136
    TERI, Bangalore 179
reflective roof, Neelam and Ashok Saigal, Gurgaon 104
renewable energy systems
    Sudha and Atam Kumar, Delhi 97
    IREP TC, Delhi 86
    Himurja, Shimla 23
    IREP TC, Delhi 89
    Mary Mathew, Bangalore 175
    Mohini Mullick, Nainital 31
    Sudha and Atam Kumar, Delhi 99
    TERI, Bangalore 180
    use of 1
    Vikas, Auroville 207
    WBPCB, Kolkata 200
    WBREDA, Kolkata 193

renewable energy technologies, RETREAT, Gurgaon 111
renewable sources of energy, RETREAT, Gurgaon 111
residence for
    Madhu and Anirudh, Panchkula 59
    Mahendra Patel, Ahmedabad 161
    Neelam and Ashok Saigal, Gurgaon 102
    Sudha and Atam Kumar, Delhi 97
    Mohini Mullick, Nainital 30
RETREAT, Gurgaon 111
roof based air-heating system 12
roof garden, TERI, Bangalore 180
roof insulation
    and water proofing 212
    Mary Mathew, Bangalore 175
    SPH, Jodhpur 168
roof treatment, WALMI, Bhopal 121

S

Sangath, Ahmedabad 151
Sarai for Tabo Gompa, Spiti 55
SEC (*see* Solar Energy Centre, Gurgaon)
sewage management, WALMI, Bhopal 122
shading devices 9
    WBPCB, Kolkata 197
shading, IREP TC, Delhi 87
Shimla
    climate in 21
    Himurja office building 21
    HP SCB 25
    MLA hostel 33
Silent Valley, Kalasa 201
site of Bidani House, Faridabad 66
site planning, Nisha, Goa 185
site, PEDA, Chandigarh 62
solar active heating systems
    MLA Hostel, Shimla 33
    HPSCB, Shimla 25
Solar Energy Centre (SEC), Gurgaon 128
solar energy, use of, Mahendra Patel, Ahmedabad 161
solar hot water system
    Sri Aurobindo Ashram, New Delhi 95
    Sudha and Atam Kumar, Delhi 99
solar passive architecture, RETREAT, Gurgaon 111
solar passive heating systems, MLA Hostel, Shimla 33
Solar Passive Hostel (SPH), Jodhpur 166
solar passive systems, HP SCB, Shimla 25
solar passive techniques, incorporation of 1
solar photovoltaic system
    IREP TC, Delhi 89
    Mahendra Patel, Ahmedabad 163

solar photovoltaics, WBPCB, Kolkata **200**
solar radiation **7, 11, 12, 15**
   controlling of, IREP TC, Delhi **87**
solar thermal evaluations, LEDeG Trainees' Hostel, Leh **52**
solar water heating system
   RETREAT, Gurgaon **113**
   Himurja, Shimla **23**
   Mahendra Patel, Ahmedabad **163**
solarium **13**
   Himurja, Shimla **22**
   LEDeG Trainees' Hostel, Leh **50**
SOS TCV (*see* SOS Tibetan Children's Village, Dehradun)
SOS Tibetan Children's Village (SOS TCV), Dehradun **76**
south orientation, Sudha and Atam Kumar, Delhi **98**
space cooling
   concept, AIIS, Gurgaon **140**
   IIHMR, Jaipur **148**
space planning, Visitors' Centre, Auroville **220**
space-controlling loads **161**
SPH (*see* Solar Passive Hostel, Jodhpur)
Spiti, Sarai for Tabo Gompa **55**
structural system, TCI Ltd, Gurgaon **74**
structure optimization, Neelam and Ashok Saigal, Gurgaon **104**
sunspaces **13**
surface-to-volume ratio **3**
sustainable
   habitat, model of, RETREAT, Gurgaon **112**
   supply of energy, RETREAT, Gurgaon **112**
   use of natural resources, RETREAT, Gurgaon **111**

T

Tapasya block, Sri Aurobindo Ashram, New Delhi **91**
TCI Ltd (*see* Transport Corporation of India Ltd, Gurgaon)
temperature, diurnal swings in, Bidani House, Faridabad **68**
TERI office building-cum-guest house, Bangalore **177**
terrace garden, AIIS, Gurgaon **139**
thermal efficiency improvement, MLA Hostel, Shimla **35**
thermal insulation **5**
thermal performance, Airport and staff housing colony, Kargil **46**
thermally comfortable design, NMC CHS, Gurgaon **135**
Torrent Research Centre (TRC), Ahmedabad **155**
traditional
   flavour, AIIS, Gurgaon **141**
   hill architecture, Mohini Mullick, Nainital **30**
Transport Corporation of India Ltd (TCI Ltd), Gurgaon **69**
TRC (*see* Torrent Research Centre)
Trombe wall **11**
tropical skylight, Neelam and Ashok Saigal, Gurgaon **104**

U

underground earth tunnels, RETREAT, Gurgaon **114**

V

ventilation
   SEC, Gurgaon **130**
   WBREDA, Kolkata **191**
   WALMI, Bhopal **121**
Vikas Apartments, Auroville **206**
Visitors' Centre, Auroville **218**

W

waste management techniques
   NMC CHS, Gurgaon **134**
   WBPCB, Kolkata **200**
   RETREAT, Gurgaon **111**
   Silent Valley, Kalasa **204**
waste material use, Sangath, Ahmedabad **153**
waste recycling, Sudha and Atam Kumar, Delhi **99**
waste water recycling
   La Cuisine Solaire, Auroville **212**
   RETREAT, Gurgaon **115**
Water and Land Management Institute (WALMI), Bhopal **119**
water
   bodies, location of **3**
   channels, Sangath, Ahmedabad **153**
   conservation, Silent Valley, Kalasa **202**
   conservation, Sudha and Atam Kumar, Delhi **99**
   management techniques, NMC CHS, Gurgaon **134**
   management, Vikas, Auroville **207**
   wall **12**
WBPCB (*see* West Bengal Pollution Control Board, Kolkata)
WBREDA (*see* West Bengal Renewable Energy Development Agency, Kolkata)
West Bengal Pollution Control Board (WBPCB), Kolkata **195**
West Bengal Renewable Energy Development Agency (WBREDA), Kolkata **189**
wind for cooling, exploitation of, IREP TC, Delhi **88**
wind tower **14**
   SPH, Jodhpur **167**
wind-driven evaporative cooler, property at Civil Lines, Delhi **83**
window design, SPH, Jodhpur **168**
windows for daylighting, SEC, Gurgaon **129**
WALMI (*see* Water and Land Management Institute, Bhopal)

A lamp that consumes 80% less electricity than an ordinary bulb and lasts six times longer! Since it is, however, a novel concept in lighting, you may well find the need for further light on the Ecotone range of Compact Fluorescent Lamps. of white and yellow light. From the pear-shaped Ambiance to the globe-shaped Decor, there's an Ecotone to match your every lighting and aesthetic need.

Where can one use Ecotones?
The Ecotone range has versatile uses- in households as well as in

ECOTONE: THE PLAIN TRUTH about THE WONDER LAMP

9W to 23W- does lesser mean dimmer?

On the contrary, the light output of a 20W Ecotone is equivalent to that of a 100W bulb. Available in a range between 9W and 23W, each Ecotone consumes one-fifth the wattage of an ordinary bulb. It is this very feature that enables it to save electricity.

A COMPARISON OF LIGHT BETWEEN CFL LAMPS AND AN ORDINARY BULB

Normal Bulb	40W	60W	75W	100W	125W
Ecotone Saver	9W	11W	15W	20W	23W
Ecotone Crystal		11W	15W	20W	
Ecotone Decor			15W	20W	
Ecotone Ambiance				16W	
Ecotone Super			15W	20W	23W

Is there only one Ecotone bulb?

Available in a variety of wattages, there is a range of shapes and sizes to choose from. With the option establishments. With different hues of light, they can create the required ambience right through your home. They make for ideal working light and bedside lamps, as they do not generate the excessive amount of heat that ordinary bulbs do. In shops and workplaces where electricity is used through the day, switching to Ecotones would reduce costs significantly.

See what you save when you switch to a Philips Ecotone of comparative wattage*

100W GLS	Ecotone 20W	Saving per year per lamp
365x100x6x2.5 / 1000	365x20x6x2.5 / 1000	
Rs. 547.50	Rs. 109.50	Rs. 438.00

* @ Rs. 2.50/- per unit at 6 burning hours per day

What about starter and fitting trouble?

Thanks to its electronic circuitry, the Ecotone starts in an instant and is flicker-free. It combines all the positive features of an ordinary bulb and a tubelight. Which is why it works efficiently within the range of 170V and 250V. This makes it ideal for Indian conditions (i.e. low voltage and severe voltage fluctuations). Moreover, it fits in a normal bulb-holder just as a normal bulb would, without an additional ballast or starter. Now that all your queries are answered, isn't it time you let the Ecotone work wonders for you?

Let's make things better. **PHILIPS**

PhL000/00

To know more about this wonder lamp contact your nearest electrical shop or write to us at Lamps Marketing, Philips India Ltd, P-65 Taratolla Rd, Calcutta 700 088

In the new millennium, switch over to a million-year-old technology

Building Integrated PV System for Mamata Machinery, Ahmedabad

Integrated Solar PV Gasifier Hybrid Power Plant for the Tata Energy Research Institute

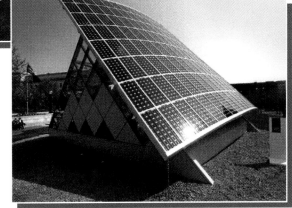

Visitors' Centre at BP Chemicals, Baglan Bay, Wales

Switch over to the sun.

For more than a decade now, Tata BP Solar has helped millions of people switch over to an environment-friendly, long-lasting and cost-effective energy source: the sun. Set up with the objective of developing and propagating Solar Power, Tata BP Solar designs, engineers, manufactures, supplies and installs a wide range of Solar Photovoltaic and Solar Thermal Systems to cater to both domestic and industrial needs.

The largest Solar Company in India, **Tata BP Solar** is a joint venture between the **House of Tata** and **BP Solar, UK** (now BP Solarex, USA, a wholly owned subsidiary of BP Amoco), the largest Solar Company in the world. A combination that spells quality all the way! An **ISO 9001** Company, Tata BP Solar has also been awarded the **ISO 14001** Certification for design, development, manufacture, installation and servicing of its SPV Systems.

So go on, light up your new millennium with solar power. With a little help from Tata BP Solar.

TATA BP SOLAR INDIA LIMITED

Plot No. 78, Electronics City, Hosur Road, Bangalore – 561 229
Tel : 080-8520083/8521016 Fax : 080-8520972/8520116
Email : tatabp@solar.ind.bp.com Website : www.tatabpsolar.com

Regional Offices :
Ahmedabad : 079-6578464. **Calcutta** : 033-2880452.
Mumbai : 022-4934426. **New Delhi** : 011-6924120.

POWER SAVERS

with COMFORT & CONVENIENCE

Rex Time Switches, from MDS Legrand, are auto-switching devices ideal for compound lighting and garden lighting in bungalows, commercial complexes and building complexes. Time Switches are to be programmed in such a way that the lighting comes on only when it is dark and switches off as soon as it becomes daylight. **This prevents wastage of electricity and thus saves power. Moreover, one can do away with manual switching to provide comfort and convenience to the user.** Also, Time Switches are used to switch on/off air conditioners in offices and bedrooms. Time Lag Switches are used with staircase lighting where power is to be kept on only for a short duration of 3 to 5 minutes.

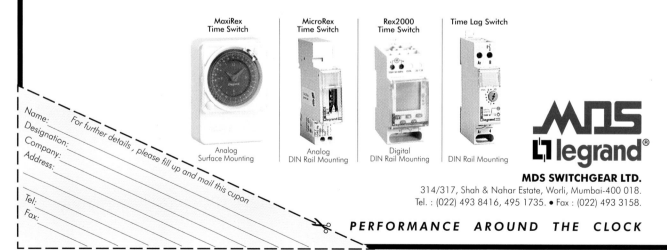

MDS SWITCHGEAR LTD.
314/317, Shah & Nahar Estate, Worli, Mumbai-400 018.
Tel. : (022) 493 8416, 495 1735. • Fax : (022) 493 3158.

PERFORMANCE AROUND THE CLOCK

BLUE STAR'S NEW ADVANCED CHILLER TECHNOLOGY BOOSTS ENERGY SAVING UPTO 75%

Millennium Centrifugal Chillers with Variable Speed Drive give you maximum energy efficiency and unprecedented savings at Real-world conditions.

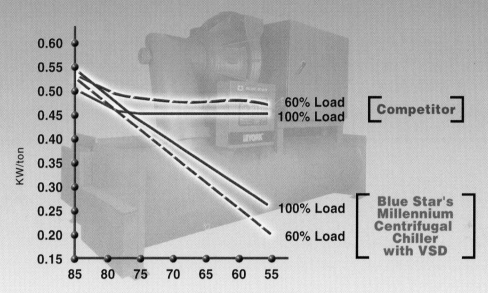

Blue Star Limited, India's largest central airconditioning company has introduced the advanced Millennium Centrifugal Chillers with optional Variable Speed Drive in technical collaboration with York International Corp, USA. Millennium Centrifugal Chillers with VSD, when working at off-design conditions, which is nearly 99% of the time, operate at energy usage of 0.40 to 0.30 and even 0.20 kW/TR, as compared to other chillers' energy usage ratio of 0.60 to 0.50 kW/TR. This, in turn, leads to an average 30% annual energy saving, and at light loads the savings can reach upto 75%.

VSD also acts as a soft starter and results in starting current less than half compared to conventional chillers, enabling use of smaller generators, resulting in substantial savings in the cost of generators and backup electrical systems.

BLUE STAR LIMITED, (Airconditioning Projects Division)

Northern Region : Block 2A, DLF Corporate Park, Qutab Enclave Phase - III, Gurgaon -122002. Tel.: 6359001-20. Fax:-0124-6359220.
Western Region : Blue Star House, 9A, Ghatkopar Link Road, SakiNaka, Mumbai-400072. Tel.: 8523600, Fax: 022-8522988.
Southern Region : 133, Kodambakkam High Road, Chennai-600034. Tel.: 8272056, Fax: 044-8253121.
Eastern Region : 7, Hare Street, Calcutta - 700001. Tel.: 2480131-33, Fax: 033-2481599.
For more information visit our website at www.bluestarindia.com

Buying an AC?
Take home one with a built-in savings account.

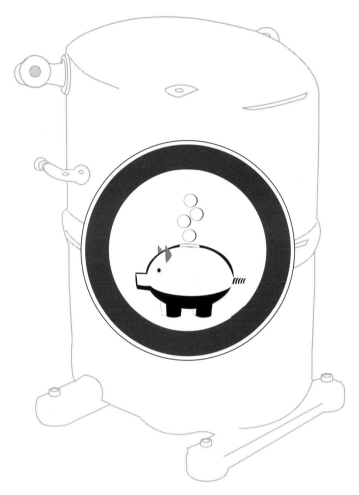

We certainly don't recommend you place currency notes in your air-conditioner. However, we do believe you should invest your hard-earned money in an air-conditioner fitted with a Power Slash CR6 compressor from Kirloskar Copeland. Remember, the compressor is the heart of your air-conditioner. Power Slash CR6 compressors are silent in operation, yet offer uncompromisingly energy-efficient performance that results in major savings on power bills.
What's more, they are built to withstand wide fluctuations in voltage. When you're buying an AC, check to see if there's a Power Slash CR6 compressor inside.
And open a new savings account.

Silent Cooling, Super Saving Compressors.
KIRLOSKAR COPELAND LIMITED
(a subsidiary of Kirloskar Brothers Ltd)

Marketing Head Office: 1202/1, Ghole Road, Pune 411 005 Tel.:020-5534988, 5534998 Fax:020-5536350

Congratulations TERI!!

We are proud to be a part of this movement of Energy Efficient Building Systems.

Janus Engineering is an electrical contracting firm with a large portfolio of very prestigious projects such as:

- RETREAT (Resource Efficient TERI Retreat for Environmental Awareness and Training) in Gurgaon;

- Biogas-based, Grid-interactive 2.7 MW power plant for SOM Distillery, Bhopal;

- Fertilizer Jetty at Port Rozi for Cargill India Ltd

- Office interior electrification for
    - Electrolux
    - National Power
    - Citi Bank
    - Electricite De France
    - Castrol India

For further information, please contact

Janus Engineering (P) Ltd.
K-55, Jangpura Extension, New Delhi – 110 014
Tel. 431 5806 • E-mail janus_eng@satyam.net.in

Confoss Constructions
Engineers, Contractors & Builders

Civil, Sanitary, Electrical, etc.

- Housing projects
- High-tech buildings
- High-rise buildings
- Industrial projects
- Interiors
- Auditoriums
- Hospitals
- Roads, culverts, landscaping and other associated civil works

Resource Efficient TERI Retreat for Environmental Awareness and Training, Gual Pahari, Gurgaon

The construction of RETREAT (Resource Efficient TERI Retreat for Environment Awareness and Training) building and technology is one of the major achievements of our organization. Our 15 years of experience in different fields of construction has made us to make a model building for TERI. The building has different features apart from normal building construction activities such as earth air tunnel 4 metres below ground for space-conditioning, insulation of external walls by thermocole, vermiculite on roof with China mosaic for insulation of roof and reduction in temperature, solar chimneys for outlets of hot air, gasifier building, and so many others.

The skill of engineering is used in construction of India's first solar roof at a height of 7 metres (high Atrium) at RETRET is one of the best achievements.

▶ 18-storeyed residential building

During the construction period, specialized tools and plants along with skilled workers and quality-minded engineers were deployed. The shuttering and casting of SPIRAL S/CASE in the lobby were typical. The civil and allied works of the root zone system (for recycling of waste water) were also carried out by us as an associate of THERMAX.

The RCC road and culvert along with typical rubble masonry done in landscaping was done with modern technology. The building has been made to the entire satisfaction of the clients.

We are thankful to TERI for creating a unique model – RETREAT – and giving us a chance to complete the project.

For further information, please contact:

Confoss Constructions
Engineers, Contractors & Builders

E-108, Greater Kailash Enclave-I, New Delhi-110048
Tel. 91-11-6465223, 91-11-6445507
Fax 91 11 6464823 • Cell 98110 70358
E-mail Confoss_mantraonline.com

Our Competition is running as fast as they can.

(When it comes to CARYAIRE, there is no competition)

At CARYAIRE, we give you the competitive edge with most reliable products in HVAC Industry.

The unmatched product range includes Air Distribution Products, Air Treatment Units, Ventilation Units, Industrial Evaporative Cooling Units, Centrifugal Fans, Inline Fans, Liquid Chillers, VAV's, Smoke & Fire Control Devices, Modular VAV Systems, Sound Attenuators, in association with the World's market leaders.

At CARYAIRE, you can be sure of getting a product to your specifications, a product manufactured with precision, a product backed by a team of highly experienced & trained Engineers & Technicians, & a product delivered to you on time.

CARYAIRE........ the industry benchmark for quality, durability & reliability.

For more information write to us.

CARYAIRE
EQUIPMENTS INDIA PVT. LTD.
The air of excellence

A-10, Sector-59, **NOIDA** - 201 301 Phones : 011-8-580553, 54, 55, 56 Fax : 011-8-580557
E-mail : caryaire@vsnl.com
WEBSITE : http://www.caryaire.com
MUMBAI Phone : 022-8852617 Fax : 022-8844471

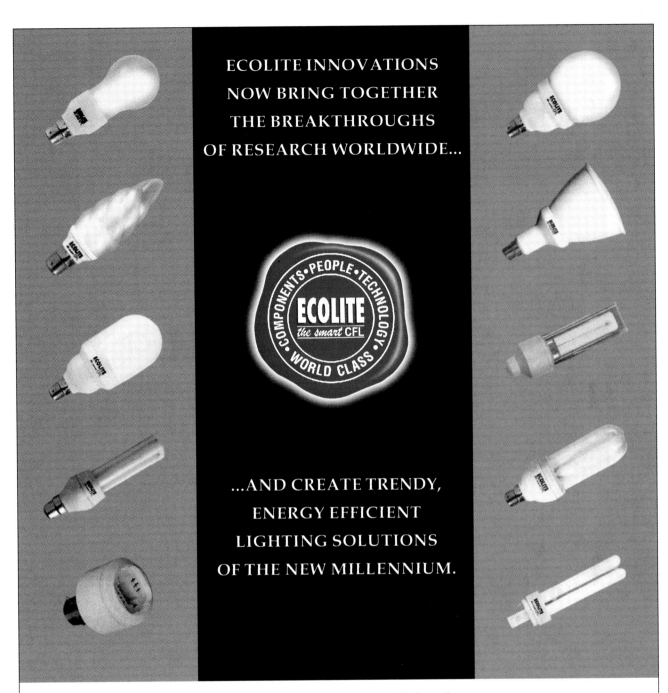

... to improve

INDOOR AIR QUALITY

without overloading the AIRCON system.

Here's the proven range of
equipment for pre-conditioning fresh air
effectively and **economically**.

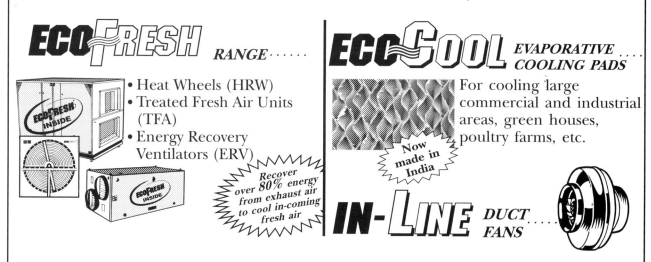

ECOFRESH *RANGE* ......

- Heat Wheels (HRW)
- Treated Fresh Air Units (TFA)
- Energy Recovery Ventilators (ERV)

Recover over 80% energy from exhaust air to cool in-coming fresh air

ECOCOOL EVAPORATIVE COOLING PADS ....

For cooling large commercial and industrial areas, green houses, poultry farms, etc.

Now made in India

IN-LINE DUCT FANS .....

We are specialists when it comes to treating fresh air, desiccant-based drying, energy recovery and cooling. Our experience in the indoor air environment business is unrivalled !

ARCTIC INDIA ENGINEERING PVT. LTD.

Delhi : 20, Rajpur Road, Delhi 110054. Ph.: 3912800. Fax: (011) 3915127. E-Mail: enquire@pahwa.com
Mumbai : 319, T.V. Industrial Estate, Worli, Mumbai 400025. Ph.: 4935155. Fax: (022) 4931020. E-Mail: aisbom@bom5.vsnl.net.in
Baroda : 326 / 327, Race Course Towers, Gotri Road, Baroda 390015. Ph.: 351493. Fax: (0265) 342080. E-Mail: aisbar@pi.zaverchad.co.in
Calcutta : Shivam Chambers, 53, Syed Amir Ali Avenue, Calcutta 700019. Ph.: 2472541. Fax: (033) 2478450. E-Mail: aiscal@cal.vsnl.net.in
Bangalore : 213 / 60, 1st Floor, 11th Cross, Wilson Garden, Bangalore 560027. Ph.: 2243411. Fax: (080) 2243412. E-Mail: shilpi@bgl.vsnl.net.in
Hyderabad : House No. 12-13-699/3/B, Street No. 20, Nagarjuna Nagar Colony, Tarnaka, Hyderabad 500007.
Ph.: 7154243. Fax: (040) 7174059. E-Mail: shilphyd@hd2.dot.net.in
Chennai : N-333, Lotus Colony, Annanagar (East), Chennai 600102. Ph.: 6287231. Fax: (044) 6203829. E-Mail: ais@md3.vsnl.net.in
Website : www.drirotors.com

A PAHWA ENTERPRISE

If these are the different parts of your building,

AIR HANDLING UNITS

CHILLERS

VENTILATION FANS

PUMPS

here's the brain.

METASYS

ELECTRICAL PANEL

Presenting METASYS, the hi-end integrated building management system from Johnson Controls. It integrates the various self-reliant systems in your building to ensure its smooth performance.

Be it the monitoring and controlling of heating, ventilating and air-conditioning, to incorporating lighting, power monitoring equipment, office automation, public address system, EPABX systems, internet connectivity, variable speed drives, refrigeration systems, METASYS, does it all. METASYS is the only system of its kind worldwide that has connectivity solutions to more than 800 different equipments. By ensuring comfort, security and safety, METASYS creates Quality Building Environments, besides reducing your energy costs.

So, do you want to equip your building with a brain?

LIFTS

GENERATOR SETS

FIRE CONTROL

JOHNSON CONTROLS

JOHNSON CONTROL (INDIA) PVT. LTD.
B-37, R. V. House, Veera Desai Road, Off Link Road, Andheri (W), Mumbai - 400 053.
PH: 91 - 22 - 636 1734. FAX: 91 - 22- 632 0343. **NEW DELHI** - PH: (011) 6149291/ 614 9289. FAX: (011) 614 9227.
CALCUTTA - PH: (033) 280 1216/ 2800253. FAX: (033) 280 1914. **BANGALORE** - PH: (080) 221 4923. FAX: (080) 224 2403.

ACCESS CONTROL

A **TATA** Enterprise

And you thought only computers could be upgraded.

Announcing the special "Upgrade Your Voltas AC" offer

VOLTAS
AIR CONDITIONERS
SPREAD HAPPINESS EVERYWHERE

Call **Voltas Crystal Care** and get an upgrade on your old Voltas air conditioner. Whether you have a split or window unit, we'll upgrade it to the latest in looks and technology. For example, we can fit a **microprocessor with a remote control, change your grille (colour options are available)** and offer you a host of other improvements. Just call **498 6666** or **372 0978** and we'll send a Voltas professional to your doorstep. What's more, while your air conditioner is in the workshop, a replacement will be provided to you free of charge. All our services are backed by a 1 year guarantee. So, call us and we'll make your air conditioner as good as new in no time.

Make sure your air-conditioner has what it needs for superior performance.

A Tecumseh compressor.

Tecumseh compressors are the choice of leading air-conditioner manufacturers, trusted for their superior cooling, energy efficiency and silent operation. You should insist on them too. Whenever you buy an air-conditioner, ask specifically for one with a Tecumseh compressor. It will give you a lifetime of cool comfort.

TECUMSEH INDIA

॥ उष्णात् मा शीतं गमय ॥
(Leading from heat to cool comfort)

Tecumseh Products India Limited, Balanagar Township, Hyderabad - 500 037. Tel: + 91 (40) 3076198, 3075880.

TECUMSEH, INDIA'S LARGEST COMPRESSOR MANUFACTURER

For centuries ceramics have been valued as works of art.

The tradition continues

Orient tiles. Truly beautiful tiles for charming homes.

If you dream of a breath taking interior, you can turn it into reality with Orient. Whose ceramic tiles allow for innovative wall and floor patterns. Orient tiles also offers sanitary ware from American Standards. Just what to need to create a bathroom of intricate beauty. From now all you need is an eye for design. And Orient tiles.

Tomorrow's trends, today

ORIENT CERAMICS AND INDUSTRIES LIMITED
Iris house, 16, Business Centre, Nangal Raya, New Delhi-110046.
Tel: 5501206, 5511274, Fax: 5511273. Mumbai: TEL:(022)6700214, 6700216, Fax: 6204411.
Kolkatta: (033)4635508, 4635509, Fax: 4635508.
Chennai: Tel:(044)6401930,6401935, Fax: 5324320,2632068